BELL NONLOCALITY

Valerio Scarani, Principal Investigator and Professor, Centre for Quantum Technologies, National University of Singapore.

Valerio Scarani was born in Milan in 1972. He graduated from Ecole Polytechnique Federale de Lausanne in 1996 and received his doctorate in physics from the same institution in 2000. He then moved to the University of Geneva, where he started working on quantum information science, notably quantum cryptography and Bell nonlocality. In 2007 he joined the National University of Singapore, as a member of the Physics Department and Principal Investigator at the Centre for Quantum Technologies.

Bell Nonlocality

Valerio Scarani

*Centre for Quantum Technologies and Department of Physics,
National University of Singapore*

OXFORD
UNIVERSITY PRESS

Great Clarendon Street, Oxford, OX2 6DP,
United Kingdom

Oxford University Press is a department of the University of Oxford.
It furthers the University's objective of excellence in research, scholarship,
and education by publishing worldwide. Oxford is a registered trade mark of
Oxford University Press in the UK and in certain other countries

First published 2019
First published in paperback 2024

Published in the United States of America by Oxford University Press
198 Madison Avenue, New York, NY 10016, United States of America

British Library Cataloguing in Publication Data
Data available

Library of Congress Cataloging in Publication Data
Data available

ISBN 978–0–19–878841–6 (Hbk.)
ISBN 978–0–19–889682–1 (Pbk.)

DOI: 10.1093/oso/9780198788416.001.0001

To my parents, Giancarlo and Angela,
and to all the others who have taught me
to open my eyes in admiration
instead of squinting in doubt.

Preface

Then I saw that all toil and all skill in work come from a man's envy of his neighbour. This also is vanity and a striving after wind.

Ecclesiastes, 4:4

Teacher: may you be eager to make your students understand quickly what has cost you hours of study to see clearly.

J. Escrivá, *Furrow*, 229

This is a monograph on Bell nonlocality. If you are reading it, you probably have already got an idea of what it is all about—at any rate, all the motivation and introduction have been packed in chapter 1 for the sake of its self-consistency, so I won't repeat them here. Let me just explain what this book tries to be, and what it is certainly not.

This book has been written with the clear awareness that *nonlocality is a phenomenon* i.e., a fact of observation. It is neither speculation, nor a notion mediated by a theory: It can be defined in its own right, and tested in its own right. The fact that quantum theory predicts those observations is a remarkable feat for that theory. The fact that the idea was arrived at by purely foundational thinking, without any application in mind, is an encouragement not to fall into strict pragmatism. But, beyond historical avatars and our current theories, the phenomenon is here to stay—just like apples will keep falling, unaware of Newton and Einstein, indifferent to whether we attribute their trajectory to a force or a curvature.

Because of this, I have opted for presenting the phenomenon from our current perspective: The reader won't find here a presentation of the historical debates or the exegesis of the masters' writings. The book focuses on the concepts and on the mathematical description of nonlocality, both *per se* and in the context of quantum theory. It is structured as follows:

Part I contains the core material (although chapters 4 and 5 could be just skimmed through in a first reading).

Part II shows the applied side[1] of nonlocality: "Device-independent" certification of devices in the context of quantum technologies.

Part III extends towards more foundational matters related to nonlocality.

The **Appendices** expand on topics that are important but would have clogged the flow of the main text.

---

[1] Which must not be read as "the dark side": If that were my stance, I would not have devoted so much of my research and a whole part of this book to it.

A famous old joke says that Italians write fifty-page pamphlets entitled *Tutto su*, while German multi-volume treatises are called *Einführung zu*. As an Italian academically formed in Switzerland,[2] I have tried to hit somewhere in the middle. In particular:

- Parts II and III were written keeping in mind that this is a book on nonlocality. Thus, part II does not try to be a manual of device-independent certification (soon, one will be needed). Similarly, part III is not to be taken as an abrégé of foundations of physics: Even staying only in the neighborhood of discussions about quantum theory, very important topics have not been mentioned.
- Experimental evidence is crucial to science, and evidence for Bell nonlocality is abundant, as unquestionable as for any other physical phenomenon. I have scattered some citations to few important milestones, but there is no chapter devoted to experiments, only a very sketchy appendix. For challenges in the lab, others are more likely than myself to provide relevant insight.
- Some topics have been left out, which could have been fitted here or there: Monogamy of nonlocality, Grothendieck constant, and inequalities with unbounded violation, random choice of measurement settings . . . References about these and other missing topics can be found in a review article that I was honored to co-author [Brunner, Cavalcanti, Pironio, Scarani, and Wehner (2014)].

Since we are speaking of a review article: This book is not one. I have made a conscious effort to give credit where it is due, notably when attributing priorities; but also, to keep the list of references down to a reasonable length. I have written under the assumption that the internet exists, and that starting from a few relevant references the reader will be quickly able to access related works.

---

[2] I know that Switzerland is not Germany, and besides I was based in Suisse romande—but at least I was North of the Alps for quite a while.

Acknowledgment

Over the years, I have had the chance to interact with many colleagues on the topic of nonlocality (those who have also provided precious feedback on drafts of this book are marked by a*).

Two people deserve a very special mention: Antoine Suarez and Nicolas Gisin*.

Antoine introduced me to this field while I was an undergraduate fascinated by chaos and fractals (already randomness, in a sense). For various reasons I embarked in a Ph.D. on Nuclear Magnetic Resonance (NMR) experiments on magnetic nanostructures:[1] While encouraging me to carry it to good completion, Antoine asked me to join him in writing down his idea of before-before experiment. Later he introduced me to Nicolas. After all these years, I am still amazed at his undimmed drive towards unity of knowledge.

Nicolas dared to hire me as a postdoc in quantum information theory, notwithstanding that completely unrelated Ph.D. topic. When hiring me, he left it clear: "*Avec la nonlocalité on ne gagne pas sa vie*". Indeed, I would not be where I am if I had not worked also in cryptography. A few years later, when in a bout of independence I told him that I would resign from his group, he had the class of staying patient, instead of firing me on the spot as I would have deserved. This gave me time to reconsider that silly decision, get back to a positive spirit, and resume work, in time to be part of the pioneering work on device-independent certification. Nicolas has just officially retired, but I am sure that neither his brain nor his computer will notice the difference.

The others ... if I were to tell all the stories, I'd have to write another book, so I'll just list them here in alphabetical order: Antonio Acín*, Mafalda Almeida, Rotem Arnon-Friedmann*, Alain Aspect, Jean-Daniel Bancal*, Jonathan Barrett, Cyril Branciard*, Gilles Brassard, Časlav Brukner, Nicolas Brunner*, Jeffrey Bub, Harry Buhrman, Francesco Buscemi*, Adán Cabello*, Daniel Cavalcanti*, Giulio Chiribella, Andrea Coladangelo*, Roger Colbeck, Daniel Collins, Michele Dall'Arno, Mauro D'Ariano, Gonzalo de la Torre, Artur Ekert*, Berge Englert, Chris Fuchs, Otfried Gühne, Michael Hall*, Lucien Hardy, Paweł Horodecki, Jed Kaniewski*, Dagomir Kaszlikowski, Adrian Kent, Barbara Kraus, Christian Kurtsiefer, José Latorre, Yeong-Cherng Liang*, Noah Linden, Alexander Ling, Norbert Lütkenhaus, Lluís Masanes, Serge Massar, Matthew McKague, André Méthot, Michele Mosca, Miguel Navascués*, Tomasz Paterek, Marcin Pawłowski, Paolo Perinotti, Stefano Pironio*, Martin Plesch, Sandu Popescu, John Preskill, Renato Renner*, Grégoire Ribordy, Ana Belén Sainz*, Nicolas Sangouard,

---

[1] *En passant*, thanks to my Ph.D. supervisor Jean-Philippe Ansermet for being always encouraging to someone who had obviously landed at the wrong place.

Rüdiger Schack, Roman Schmied, Yaoyun Shi, Tony Short, Jamie Sikora, Christoph Simon, Paul Skrzypczyk*, Rob Spekkens, André Stefanov, Sébastien Tanzilli, Nicholas Teh, Wolfgang Tittel, Philipp Treutlein, Antonios Varvitsiotis*, Umesh Vazirani, Tamás Vértesi, Thomas Vidick*, Paolo Villoresi, Stephanie Wehner, Reinhard Werner, Howard Wiseman, Andreas Winter, Michael Wolf, Stefan Wolf, Elie Wolfe, Henry Yuen, Hugo Zbinden, Anton Zeilinger, Marek Żukowski.

Since I built my own research group, I have worked with many excellent students and researchers. I proudly keep the full list in the group's webpage. Those who deserve to be mentioned in relation with nonlocality are, again in alphabetical order:[2] Charles-Edouard Bardyn, Yu Cai, Wan Cong, Benjamin Goh, Koon Tong Goh, Yunguang Han, Asaph Ho, Melvyn Ho, Yun Zhi Law, Thinh Le, Zhi Xian Lee, Xinhui Li, Charles Lim, Rafael Rabelo, Lana Sheridan, Ernest Tan, Yukun Wang, Xingyao Wu, Tzyh Haur Yang, and Yuqian Zhou. When finishing this book, I decided to put it to the test and proposed a graduate module based on it. Among the students who took it, special thanks to Adrian Nugraha Utama, Ignatius William Primaatmaja, Angeline Shu, and Noah Van Horne for feedback and discussions.

After speaking of the future, a glance to the past. Among my many teachers, I want to single out Alberto Torresani, who taught me philosophy in Liceo Scientifico Argonne, and François Reuse, whose opened my eyes to a proper understanding of quantum theory during my years in Ecole Polytechnique Fédérale de Lausanne (EPFL).

My family and closest friends will hardly understand the title of this book, let alone its content. But without their support, I would not have been able to complete the task.

---

[2] Charles and Rafael have since becomes colleagues, but are mentioned in this list because they started in my group. This is also the reason why Daniel and Jean-Daniel were mentioned previously: When they came to work with me, they had already acquired a great competence in nonlocality.

Contents

Part I

Classic Bell Nonlocality

This first part deals with Bell nonlocality *per se*, following a synthetic rather than a historical outline. The first two chapters deal with the definition of Bell nonlocality, the conditions under which it can be demonstrated, and its mathematical formalization. The description of the phenomenon within quantum theory is then introduced, leading to a few general results and a large number of examples.

1

First Encounter with Bell Nonlocality

Misunderstandings of Bell's theorem happen so fast that they violate locality.

R. Munroe, *XKCD*

This chapter serves both as an introduction to the book, and as a self-contained first presentation of Bell nonlocality.

1.1 Three Roles for Bell Nonlocality

Few scientific statements are more radical than one of the core tenets of quantum physics: *There is indeterminacy in nature*. It has accompanied quantum theory since its earliest moments: Only a few months after Heisenberg and Schrödinger independently defined the definitive formalism, Max Born suggested that the laws of the new theory should be seen as intrinsically statistical. This was to become the orthodox view. Sensing the danger, Einstein quickly wrote to Born his conviction that a theory with statistical laws could only be a temporary fix, and that determinism should ultimately be recovered. The debate continued for decades with a few flares, notably the celebrated EPR paper (Einstein Podolsky, and Rosen, 1935) and Bohr's immediate reply, but in an atmosphere of overall indifference among physicists at large. In those years, the excitement about quantum theory was not found in debating its meaning, but in its almost boundless predictive power. It has become commonplace to refer to the attitude of those years by Mermin's dictum "shut up and calculate."

Ultimately, the statistical language became the standard to which generations of physicists conformed out of inertia. If asked for evidence of indeterminacy, still today many would refer to Heisenberg's uncertainty relations, that however can only voice for indeterminacy in quantum theory, not in nature (see Appendix A.1). This is surprising because direct evidence has been compelling since 1964, thanks to the work of John Bell (Bell, 1964). He showed that the possibility of recovering a deterministic model is amenable to experimental falsification, through the observation of a phenomenon that we shall call *Bell nonlocality*. In a first approach, Bell's argument is mathematically simple (see sections 1.3–1.4); because of its importance, it has been submitted to a thorough scrutiny, from which it has emerged unscathed and actually strengthened by more solid foundations (see section 1.5 and chapter 2).

Bell Nonlocality. Valerio Scarani. © Valerio Scarani 2019. Published in 2019 by Oxford University Press.
DOI: 10.1093/oso/9780198788416.001.0001

In 1964 there was already a huge amount of experimental evidence supporting the validity of quantum theory. Nevertheless, none of those data could be used to check Bell's criterion. Dedicated experiments had to be designed. The work of Alain Aspect and coworkers is credited as the first conclusive evidence of Bell nonlocality (Aspect *et al.* 1982*b*; Aspect *et al.* 1982*a*). The evidence has been steadily growing since then; eventually, three independent experiments reported in 2015 are considered definitive (Hensen *et al.*, 2015; Giustina *et al.*, 2015; Shalm *et al.*, 2015). The main text of this book does not describe experiments; to facilitate reading the experimental literature, a quick guide is provided as Appendix B.

Discovered thanks to quantum theory, indeterminacy has been vindicated as a physical fact, independent of the theory itself. It can be circumvented only at the price of adopting even more radical postures about physics and nature themselves (see section 1.6). *This direct vindication of indeterminism is the original motivation of Bell nonlocality.* For a few decades, it was held to be its sole role too. Those who, for various reasons, were already won to the indeterministic cause had taken note of it and moved on. A series of works that started around 2005 have uncovered a second role: *Bell nonlocality provides the most compelling certification of the correct functioning of some quantum devices*, like those required to perform quantum cryptography and quantum computation. The fabrication of these devices and the development of certification tools based on nonlocality still constitute technical challenges, but we'll have to get there. Far from being an exercise in scientific archaeology, this book contains material that future quantum engineers will have to master—*in nuce* at least, this is a treatise in applied physics.

Finally, as a phenomenon independent of quantum theory, Bell nonlocality is not merely an instrument for a negative task (falsifying determinism): It has a right to citizenship in physics. As its third role, *Bell nonlocality can be used as a principle constraining possible candidates for physical theories.* Barring a few pioneering insights, this approach was also started after the year 2000. It has already contributed several new ideas and notions to the field of foundations of physics but is still very open to future developments.

These three roles of Bell nonlocality—evidence for indeterminism, certification tool for devices, and guideline for foundations—correspond to the three parts into which this book is divided, but pervade the whole text. With them in mind, we can enter the core of the subject.

1.2 Introducing Bell Nonlocality

1.2.1 Setting Bell tests: Laboratories and games

Tests of Bell nonlocality, or *Bell tests* for short, are currently *experimental setups in physics laboratories*. For some years, theorists have rather chosen to present Bell tests as *games* that bear some analogy with TV quizzes, polls, exams, judicial trials, and other familiar situations.[1] The game setting is definitely better to bring up the essence of nonlocality,

---

[1] The reader may come back to this list after becoming familiar with Bell test, to find analogies and differences. For instance, in exams, the content of the answer matters, and the verifier will evaluate the

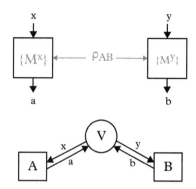

Figure 1.1 *Sketch of a Bell test for two players (the generalization to more players is straight-forward) in the* laboratory *setting (top) and in the* game *setting (bottom). After having agreed on a process for that round, each player receives an input and has to provide an output; the data of several rounds are then sorted to establish the correlations between the outputs a and b for any pair (x, y) of inputs. In the laboratory setting, we give in grey the usual representation of the process in quantum theory: A quantum state ρ is prepared and some measurements are chosen; which measurement is actually performed in each round is determined by the input. None of this enters the definition of Bell nonlocality. In the game setting, we introduce a verifier V that queries the players Alice and Bob and collects their answers.*

and we shall mostly follow it in this book. Nonetheless, as we shall also see, many important discussions cannot be fully appreciated without going back to the lab. The two settings are sketched and compared in Figure 1.1.

In a Bell test game, the *players*, referred to alphabetically as Alice, Bob, Charlie, etc., are all on the same team. The game consists of many *rounds*. In each round, the players will be separated: Each will receive a query (*input*) and will have to provide an answer (*output*). It is useful to think of a *verifier* distributing the inputs[2] and collecting the outputs.

The rules of the game and the list of possible queries are known in advance. The players are allowed to prepare a common *strategy* before the game, which consists in deciding which *process* they will use in each round of the game. We shall also speak of the *resources* that are used in these processes. If the players were allowed to communicate among each other during the game, they would actually not be separated and could easily win any game of this type: The most powerful resources are *signaling* ones. The case is more interesting with *no-signaling resources*. The most elementary example of a no-signaling resource is a list of pre-determined outputs, one for each possible input (that is, the process consists in producing the output by reading the list). It is no-signaling

---

performance of each player without any concern for correlations. In judicial trials, the players' goal is to provide a consistent version of the story (which may not be the truth), but they don't know in advance the set of questions that they may be asked; etc.

[2] In actual experiments, the inputs are usually generated at each player's location by a "random number generator": The image of the verifier allows us to postpone the delicate discussion about randomness and its generation with physical means (subsection 1.5.3).

because, if Alice does something to her list, the other players obviously won't notice anything. In other words, Alice can't send a message to others by manipulating her list.

For instance, consider three games, each defined by one of the following rules:

(i)   The players must produce the same answer if they receive the same query.

(ii) The players must produce the same answer if and only if they receive the same query.

(iii) The players must produce different answers if both receive query "1," the same answer otherwise.

A game based on rule (i) is trivially won by the players agreeing on a fixed common output. A game based on rule number (ii) can be similarly won by agreeing on a pre-determined output for each input, provided that the number of inputs is not larger than the number of outputs. If there were more inputs than outputs, the game cannot be won with a list of pre-determined answers. Finally, no strategy based on pre-determined answers can win a game based on rule (iii).

1.2.2   The definition of Bell nonlocality

Bell locality means that *the process by which each player generates the output does not take into account the other player's input.* In other words, all correlations between the players' outputs is due to the shared resource, on whose nature no assumption is made: It can be anything, from a list of numbers on a piece of paper to two jointly programmed quantum computers. When Bell locality does not hold we speak of *Bell nonlocality*.

This notion of locality can be formalized as follows. Denote by λ the process. It does not need to be deterministic, so we can say that Alice generates a by sampling from a probability distribution $P_\lambda(a|x)$. What is crucial is that this does not take Bob's input y into account. Similarly, Bob generates b locally by sampling from a probability distribution $P_\lambda(b|y)$. If this is the case, the statistics observed by the verifier (who is not privy to λ) will be described by

$$P(a,b|x,y) = \int d\lambda Q(\lambda) P_\lambda(a|x) P_\lambda(b|y) \qquad (1.1)$$

where $Q(\lambda)$ is the probability distribution that describes the strategy, i.e., how often a specific process λ is used. Bell locality is clearly a restriction: Not all conceivable $P(a,b|x,y)$ can be written in this form. The extreme counterexample is a strategy that wins the game in which each player is supposed to output the other player's input. Also, any winning strategy for the game based on rule (iii) requires one of the players to sample from a distribution that depends on the other player's input.

Statistics will be called *local* if they can be written in the form (1.1), *nonlocal* if they cannot. A *Bell test* is a game whose winning strategy is described by nonlocal statistics.

1.2.3 On resources and semantics

If nonlocal statistics are observed, the verifier knows that the players have shared a *nonlocal resource*. If local statistics are observed, we can't say much about the resource: The players might have shared a potentially nonlocal one but have used it poorly. This sounds like elementary logic, but it triggers two crucial remarks:

- The definition of Bell locality does not rely on a prior characterization of the class of "local resources"; even less one needs to assume that there *exist* in nature resources that are intrinsically local in this sense.[3] The opposite is the case: From the definition of Bell locality, that stands its ground, one can *define* "local resources" as hypothetical resources that could only lead to local statistics. The traditional name for such local resources is *"local hidden variables" (LHVs)* or simply *"local variables" (LVs)*, the word "hidden" being a relic of the discussions on quantum theory.

- We'll see in chapter 2 that every local statistic (1.1) can even be realized with a strategy based on pre-determined outputs. This result, known as Fine's theorem, is the basis for the mathematical tools of the field. But we cannot infer from this theorem that all observed local behaviors are actually generated with pre-determined outputs, nor that the definition of Bell locality assumes determinism.

Now, were it not for quantum theory, the definition of Bell nonlocality would sound both uninteresting and uncontroversial: Nonlocal resources would be communication devices. Quantum theory[4] however forces us to enlarge, at least in principle, the list of possible nonlocal resources. Let us then consider that the players *share physical systems* in a state that quantum theory describes as ρ_{AB}, and let's assume that the process that produces the outputs is *performing local measurements* on this state. Specifically, upon receiving her input x, Alice performs a measurement on her system, with the output a of that measurement is associated to the positive operator Π_a^x. Bob acts similarly. After several rounds, all played with this process, quantum theory predicts that the statistics collected by the verifier are given by

$$P(a, b|x, y) = \mathrm{Tr}(\Pi_a^x \otimes \Pi_b^y \rho_{AB}). \tag{1.2}$$

In general, these statistics cannot be cast in the form (1.1): This is the content of *Bell's theorem* (Bell, 1964). Explicit examples will fill this book, but for the time being let us accept that *some shared quantum states are nonlocal resources*. However, it is also well

---

[3] In the same vein, notwithstanding the frequent replacement of "local" with "classical" in the field's jargon, the definition of Bell nonlocality does not rely on a definition of classicality, and even less on assuming the existence of intrinsically classical physical systems. Overall, we shall avoid speaking of classical/quantum systems or phenomena. It is correct to speak of classical theory and quantum theory, because these are well-defined (see Appendix C.1.1 for the essentials). It is also customary to speak of classical/quantum *information* to refer to the resources, insofar as described within each theory, but we won't do it.

[4] Familiarity with elementary quantum theory is given for granted in this book; more advanced topics and specific aspects of quantum information theory are summarized in Appendix C.

known[5] that shared quantum states *are not communication channels*: By acting only on her system, Alice cannot learn anything about what Bob has done with his—he could have measured it, kept it, discarded it, and Alice does not see any change in her statistics. In this sense, quantum states are *no-signaling resources* just as shared lists of numbers. *Bell nonlocality is interesting and intriguing because it can be demonstrated by sharing no-signaling resources.*

Or can it? Famously (or notoriously), quantum theory does not provide any recipe for the generation of each round's output. Would it be possible that what quantum theory describes as no-signaling resources are actually signaling ones? Einstein dubbed this possibility "spooky action at a distance". As we shall see in section 1.6, it is one possible interpretation. Among those who oppose it, some think that the wording "nonlocality" evokes too closely this unwelcome interpretation. What we called "locality" in subsection 1.2.2, they'd rather call *local realism* or *local causality*. These are elegant expressions with philosophical appeal: They remind us that we are not merely dealing with operations and observations, but with a prejudice in our *Weltanschauung* that has been shattered. However, they are also not exempt from the danger of being over-interpreted.[6]

With all their potential limitations, the wordings "nonlocality," "local realism," and "local (hidden) variables" have already enjoyed a few decades of tradition and are most probably here to stay. I hope I have said enough to prevent their misuse, and I shall use them freely. As for interpretations, we shall return to them in section 1.6.

1.3   My First Bell Test: Clauser-Horne-Shimony-Holt (CHSH)

To put these general considerations on concrete grounds, we proceed to describe some specific examples of Bell tests. For this introductory chapter, I have chosen to present five classic examples: One in this section and four in the next; several others will be presented later in the book. An elementary proof that each is indeed a Bell test is given, exploiting Fine's theorem (subsection 2.3.3) that allows considering only strategies based on pre-established answers. The relevance of each test for the certification of quantum entanglement is merely stated, leaving all the calculations for chapters 3–5.

---

[5] Even if this should be elementary knowledge, given the centrality of the claim for the content of this book, the explicit proof is given in Appendix C.1.3.

[6] It has become commonplace to split the prejudice of "local realism/causality" into two separate prejudices, "locality" and "realism" (or "causality"). To be at peace with the fact of a violation, it would then be enough to abandon either. *Abandoning locality* may legitimately mean signaling: In (1.1), one would have $P(a|x, y, \lambda)$, $P(b|x, y, \lambda)$, or both; and this modification is indeed sufficient to generate Bell nonlocality. But the meaning of *abandoning realism/causality* is by far less clear (Norsen, 2007; Gisin, 2012). It cannot mean "abandoning determinism": Determinism is not assumed in (1.1), so abandoning it does not generate Bell nonlocality. Neither should it mean "abandoning any connection with reality," reducing physics to unfounded speculations: If we abandon "local realism" it's because we accept the verdict of observation. Probably, "abandoning realism/causality" is a way of saying that only statistics are speakable, see subsection 1.6.3. But then, this alternative is at a different level than signaling: It is not a mechanism, but the statement that no mechanism should be looked for.

It should be obvious that a Bell test requires *at least two players*, otherwise there is no notion of locality. For each player, there must be *at least two possible inputs*: If some players could only receive one query, those inputs would be known to the other players. Finally, for each input, there must be *at least two possible values for the output*. We start with this simplest scenario.

The inputs of Alice are labeled $x \in \{0, 1\}$, her outputs $a_x \in \{-1, +1\}$ (labeling is of course arbitrary, this choice is convenient for the calculation to come). The inputs of Bob are labeled $y \in \{0, 1\}$, his outputs $b_y \in \{-1, +1\}$.

The rule of the game prescribes that Alice and Bob should aim at giving the same answer whenever $(x, y) \in (0, 0), (0, 1), (1, 0)$, but opposite answers when $(x, y) = (1, 1)$. We consider the score

$$S = \langle a_0 b_0 \rangle + \langle a_0 b_1 \rangle + \langle a_1 b_0 \rangle - \langle a_1 b_1 \rangle \tag{1.3}$$

where the average is taken over an arbitrarily large number of rounds. The maximal score is obviously $S = 4$.

To prove that this game is a Bell test, we need to find what score can be achieved with local resources. Invoking Fine's theorem, it is enough to see what happens when Alice and Bob have shared a pre-determined quadruple $(a_0, a_1; b_0, b_1)$ in each round. The existence of these four numbers entails the existence of a well-defined value for the derived quantity $s = a_0 b_0 + a_0 b_1 + a_1 b_0 - a_1 b_1$. Since the average of a sum is the sum of the averages, $S = \langle s \rangle$ holds. Now, for every quadruple, either $s = +2$ or $s = -2$. This is readily seen by rewriting $s = a_0 (b_0 + b_1) + a_1 (b_0 - b_1)$: Indeed, if $b_0 = b_1$ the second term is zero, if $b_0 = -b_1$ the first term is zero. In Table 1.1 we list explicitly all sixteen possibilities: Eight of them give $s = +2$ and the other eight $s = -2$. Notice how three out of four pairs of inputs contribute in the same way, but the last pair pulls the sum down (or up if it was negative). This observation will become handy later.

In each round, the verifier sees only the pair (a_x, b_y) corresponding to the inputs (x, y) he has sent, so he cannot estimate s. However, by performing statistics conditional on each pair of inputs, he can estimate the four $\langle a_x b_y \rangle$ and obtain S. Now, if $S = \langle s \rangle$ and each instance of s can only take the values ± 2, it follows that

$$|S| \overset{LV}{\leq} 2. \tag{1.4}$$

Suppose now that the verifier finds $S > 2$: He will have to admit that the players were not sharing pre-established quadruples, nor any resource that can be simulated with them— in other words, he will have to admit that *they share a nonlocal resource*.

This Bell test is called CHSH from (Clauser, Horne, Shimony, and Holt, 1969). Notice how the mathematical expression of the test is an *inequality*, here Eq. (1.4). The players will convince the verifier that they have a nonlocal resource if they manage to *violate the inequality*. John Bell's original inequality (Bell, 1964) is basically the same as (1.4), but derived under the assumption that one of the $\langle a_x b_y \rangle$ is exactly equal to 1 (Appendix A.4). Since perfect correlations can be predicted but cannot be observed, that inequality

Table 1.1 *All sixteen quadruples of pre-established values, and derivative quantities, for the CHSH test. In boldface, the term that pulls the sum s in the opposite direction as the other three. Notice that the second half of the table is the mirror image of the first half, since flipping all the signs does not change the products.*

$a_0, a_1; b_0, b_1$	$a_0 b_0$	$a_0 b_1$	$a_1 b_0$	$a_1 b_1$	s
$+1, +1; +1, +1$	$+1$	$+1$	$+1$	$+1$	$+2$
$+1, +1; +1, -1$	$+1$	-1	$+1$	-1	$+2$
$+1, +1; -1, +1$	-1	$+1$	-1	$+1$	-2
$+1, +1; -1, -1$	-1	-1	-1	-1	-2
$+1, -1; +1, +1$	$+1$	$+1$	-1	$\mathbf{-1}$	$+2$
$+1, -1; +1, -1$	$+1$	-1	-1	$+1$	-2
$+1, -1; -1, +1$	-1	$+1$	$+1$	-1	$+2$
$+1, -1; -1, -1$	-1	-1	$+1$	$\mathbf{+1}$	-2
$-1, +1; +1, +1$	-1	-1	$+1$	$\mathbf{+1}$	-2
$-1, +1; +1, -1$	-1	$+1$	$+1$	-1	$+2$
$-1, +1; -1, +1$	$+1$	-1	-1	$+1$	-2
$-1, +1; -1, -1$	$+1$	$+1$	-1	$\mathbf{-1}$	$+2$
$-1, -1; +1, +1$	-1	-1	-1	-1	-2
$-1, -1; +1, -1$	-1	$+1$	-1	$+1$	-2
$-1, -1; -1, +1$	$+1$	-1	$+1$	-1	$+2$
$-1, -1; -1, -1$	$+1$	$+1$	$+1$	$+1$	$+2$

is sufficient to prove that quantum theory predicts nonlocal resources but is untestable in experiments.

The CHSH test is the workhorse of the field, we'll study it in great detail. Let's just mention that the maximal score $S = 4$ can be reached of course by signaling,[7] but also with a hypothetical no-signaling resource called a *PR-box* (Popescu and Rohrlich, 1994) that we'll encounter in chapter 9. However, PR-boxes exist in mathematics but don't seem to exist in nature: With quantum entanglement, the maximal score is $S = 2\sqrt{2} \approx 2.8284$.

---

[7] Though rather obvious, here is one possible way in which the players can score $S = 4$ with signaling. Alice and Bob agree on a bit $a = b$. Upon being queried, Alice outputs a and sends x to Bob; Bob outputs $(-1)^{xy}b$. Notice that Alice's answer is pre-determined, but Bob's is not: He has to wait for x before producing it. So, the conclusion that both outputs could not have been pre-determined holds.

This value can be achieved by suitable measurements on the maximally entangled state of two qubits[8]

$$\left|\Phi^+\right\rangle = \frac{1}{\sqrt{2}}\left(\left|0\right\rangle\left|0\right\rangle + \left|1\right\rangle\left|1\right\rangle\right), \tag{1.5}$$

and in fact, in a sense to be made precise in chapter 7, *only* by that state and those measurements.

1.4 Four more Classic Bell Tests

This section introduces four other examples of Bell tests that should help gaining further familiarity with these notions.

1.4.1 Mermin's outreach criterion

In his effort to explain Bell nonlocality in a simple way, David Mermin (1981) conceived a Bell test that has become popular. There are two players, each with three inputs $(x, y \in \{1, 2, 3\})$ and two outputs $(a, b \in \{+1, -1\})$. Let's assume that the observation shows that $a_i = b_i$ in all cases where the same input was chosen. We are interested in $O = \sum_{x,y} P(a_x = b_y) = 3 + \sum_{x\neq y} P(a_x = b_y)$. If the outputs are pre-determined, $(a_1, a_2, a_3) = (b_1, b_2, b_3)$ can take eight values, namely $(+1, +1, +1)$, $(+1, +1, -1)$, etc. For $(+1, +1, +1)$ and $(-1, -1, -1)$, one finds $O = 9$; the other six triples give $O = 5$. Thus, certainly $O \geq 5$ for local resources. Quantum theory predicts that one can go down to $O = 4.5$. In particular, not even with quantum resources one can win perfectly the game based on rule (ii) defined in subsection 1.2.1, because that would correspond to $O = 3$.

The inequality $O \geq 5$ defines a Bell test only if $a_i = b_i$. If this assumption is removed, the inequality may be violated with LVs. For instance, the choice of pre-determined outputs $(a_1, a_2, a_3) = (+1, +1, +1)$ and $(b_1, b_2, b_3) = (-1, -1, -1)$ gives $O = 0$. This is the same weakness of Bell's original criterion, which had to be transformed into CHSH to become robust. Robust Bell tests for two players, three inputs and two outputs will be discussed in section 4.2.

1.4.2 Greenberger-Horne-Zeilinger (GHZ) test

The original nonlocality test by Daniel Greenberger, Michael Horne, and Anton Zeilinger (1989) considers four players, but the argument can be made for any number

---

[8] The notation $\left|0\right\rangle\left|0\right\rangle$ stand for $\left|0\right\rangle \otimes \left|0\right\rangle$. Here and in the rest of the book, the tensor product symbol is usually implicit when writing quantum states, while it is often explicit when writing operators. I find that this is the choice that facilitates the reading.

of players larger than two. Nowadays, we refer to the three-players version as the "GHZ test" without further qualifiers (Mermin, 1990b); this is the one we present here. As in the CHSH case, each player has two inputs and two outputs, those of Charlie being labeled z and c. Assume now that the verifier observes the following perfect correlations:

$$\langle a_0 b_0 c_1 \rangle = \langle a_0 b_1 c_0 \rangle = \langle a_1 b_0 c_0 \rangle = +1. \tag{1.6}$$

Then,

$$\langle a_1 b_1 c_1 \rangle \overset{LV}{=} +1. \tag{1.7}$$

Indeed, $\langle a_x b_y c_z \rangle = +1$ means that $a_x b_y c_z = +1$ in each single round. If the outputs are pre-determined, they must be such that $a_0 b_0 c_1 = a_0 b_1 c_0 = a_1 b_0 c_0 = +1$ in each round. By multiplying the three conditions and noticing that $a_0^2 = b_0^2 = c_0^2 = 1$, we get $a_1 b_1 c_1 = +1$. In quantum theory, however, one can have the correlations (1.6) alongside

$$\langle a_1 b_1 c_1 \rangle = -1. \tag{1.8}$$

This is obtained with suitable measurements on the state

$$|GHZ\rangle = \frac{1}{\sqrt{2}} \left(|0\rangle\, |0\rangle\, |0\rangle + |1\rangle\, |1\rangle\, |1\rangle \right), \tag{1.9}$$

and, just as for CHSH, this quantum realization is unique (see chapter 7).

As presented, the GHZ test relies on the observation of perfect correlations or anti-correlations. To cope with unavoidable imperfect situations, its score can be turned into the so-called *Mermin inequality*:

$$M = \langle a_0 b_0 c_1 \rangle + \langle a_0 b_1 c_0 \rangle + \langle a_1 b_0 c_0 \rangle - \langle a_1 b_1 c_1 \rangle \overset{LV}{\leq} 2. \tag{1.10}$$

The proof that inequality (1.10) holds is left as Exercise 1.1. Thus, contrary to what happens with CHSH, the maximal score $M = 4$ of the Mermin inequality can in principle be attained in quantum theory. We shall complete the study of this inequality and its generalizations for more parties in section 5.2.

1.4.3 Hardy's test

At this point, one may ask if one can build a Bell test for *two* players on extreme correlations that quantum resources can in principle achieve. There are indeed such examples. The first one was found by Lucien Hardy (1992; 1993). The extreme probabilities that are enforced are:

$$P(a_0 = +1, b_0 = +1) = 0 \tag{1.11}$$

$$P(a_0 = -1, b_1 = +1) = 0 \tag{1.12}$$

$$P(a_1 = +1, b_0 = -1) = 0. \tag{1.13}$$

The following inferences are then obvious:

- From (1.12): $b_1 = +1$ implies $a_0 = +1$
- From (1.11): $a_0 = +1$ implies $b_0 = -1$
- From (1.13): $b_0 = -1$ implies $a_1 = -1$.

By enchaining these three inferences we are led to a fourth one, namely "$b_1 = +1$ implies $a_1 = -1$," and thus in particular to the prediction

$$P(a_1 = +1, b_1 = +1) \overset{LV}{=} 0. \tag{1.14}$$

Another way of reaching the same conclusion is suggested in Exercise 1.2.

In subsection 4.4.1, we shall see that in quantum theory one may have the three constraints (1.11)–(1.13) and nonetheless $P(a_1 = +1, b_1 = +1) > 0$ with a maximal value of approximately 0.11. This looks absurd, since enchaining the bullet pointed three inferences looks innocuous—but it is not: Just as in the derivation of the CHSH inequality, the enchaining assumes that one can speak of both a_0 and a_1, and of both b_0 and b_1. Rigorously, the first inference should read: If $b = +1$ was found for $y = 1$, we know that $a = +1$ will be found *if the input $x = 0$ is called*; and similarly for the others.

1.4.4 The Magic Square

The rules of the Magic Square test[9] (Cabello, 2001*b*; Cabello, 2001*a*; Aravind, 2002) are slightly more complex. The test has three possible inputs per player, $x, y \in \{1, 2, 3\}$; the players are asked to output three bits each: $a_x = (a_x^1, a_x^2, a_x^3)$, $b_y = (b_y^1, b_y^2, b_y^3)$. These outputs should ideally satisfy the following conditions:

$$\prod_j a_x^j = +1, \prod_k b_y^k = -1, \text{ and } a_x^{j=y} = b_y^{k=x}. \tag{1.15}$$

To see that these conditions are impossible to fulfil perfectly with pre-determined outputs, let us arrange the nine pre-determined bits in a 3×3 square. Upon being queried, Alice outputs the x-th line of her square, and Bob the y-th column of his. The third condition says that the value at the intersection should be the same for every call (x, y) of the verifier, which means that the squares must be identical. However, the first condition

---

[9] The square made its first appearance as a test for single-player "contextuality" (see Appendix D.3 for this notion) in works of N. David Mermin and Asher Peres. Because of this, it is often called Mermin-Peres Magic Square. Here I cite the subsequent works that introduced it as a two-player nonlocality test.

implies $\prod_{x,j} a_x^j = +1$, the second $\prod_{y,k} b_y^k = -1$. If these are to be enforced, Alice's and Bob's 9-bit squares should differ in at least in one bit—and then, if the verifier calls precisely those inputs, he will see that the third condition fails.

All in all, with pre-determined outputs, Alice and Bob can satisfy the three conditions (1.15) for at most 8/9 of the rounds on average; but there exists a quantum state and measurements that can fulfil them perfectly, as we shall show in subsection 4.4.2.

1.5   A Closer Scrutiny: Addressing Loopholes

As we have just seen, Bell tests can be described in very elementary terms. But is this not too elementary, especially given the strong conclusions that are reached? Over the years, Bell tests have been submitted to tight scrutiny, searching for flaws, or *loopholes*, in the reasoning or in the implementations.

The four possible loopholes that have been identified turn out to be very different from each other: Some are mere technical fixes (that have been fixed), others border on philosophy and can be closed only under reasonable assumptions (in other words, there is a price to pay if one wants to believe that they are still open). We review them here in this order; their working will be illustrated with the CHSH test.

1.5.1   The "memory loophole," or doing proper statistics

The memory loophole is related to statistics. Basic statistics assumes that rounds of a test are *independent and identically distributed (i.i.d.)*, but this i.i.d. assumption is obviously unwarranted when it comes to such fundamental tests. Would it be possible for the players to give a false positive in a Bell test, i.e., violate a Bell inequality with local resources, by adopting a non-i.i.d. strategy, that is, by choosing the process to be used in round r based on all that has happened in the previous rounds? The answer is no.

To appreciate why, consider one round of the CHSH test. The players must choose the pre-established quadruple of values to be used in that round. They can base their choice on whatever piece of information from the past: There will always be one pair of inputs which pulls the sum in the wrong direction (see Table 1.1), and the verifier might have picked precisely that pair; so $|S| \leq 2$ still holds. This simple argument shows that there is only one way for the players to generate a false positive: Avoid the wrong pair of inputs, either by refusing to answer or by colluding with the verifier. These are, respectively, the fair-sampling loophole and the free-will loophole to be described.

Thus, even if we initially derived our Bell inequality thinking in i.i.d. terms, we have proved that *there is no memory loophole* if the Bell test is infinitely long and the verifier can extract perfect statistics. In a real test, when only finitely many rounds are possible, the Bell test must be phrased as hypothesis testing: How likely is the observed string of outputs assuming that the players are using a local resource? For such likelihood bounds, non-i.i.d. estimators must indeed be used instead of the familiar i.i.d.-based Gaussian standard deviations. We'll get back to this point in section 2.6.

1.5.2   The "detection loophole(s)," or the dangers of post-selection

If in an exam the students were allowed to decline to answer till they are asked a question of their liking, the average score would be certainly increased. The same happens for nonlocality. In the CHSH test played with local strategies, we have seen that only one pair of inputs (x, y) pulls the value of S in the wrong direction: If the players were allowed to decline answering when they receive that specific query, the verifier would never catch them at fault. There is only a subtlety: Neither of the players knows the pair (x, y), so the decision to decline must be made locally, based on x or on y alone. A possible strategy in which Alice answers always while Bob is in charge of declining is given in Table 1.2. We refer to Exercise 1.3 for a thorough study of this strategy, and to Appendix B.3 for a more rigorous quantitative approach.

The fix for this loophole is clear: *The verifier must elicit an answer in every round.* This does not sound to be a big deal, but it may be. Let us look at this loophole from the perspective of an honest experimentalist who possesses a very good nonlocal resource (say, a source of photons entangled in polarization) and very accurate measurement devices, but whose detectors have poor efficiency. If she is obliged to produce outputs in every round, in most of the rounds she'll have to produce a dummy output because the detectors won't have fired. Nonlocality is quickly washed down, and likely the observed data will be compatible with a local resource.

The situation is all the more annoying because this loophole has a very conspiratorial character. As John Bell himself stressed and many after him, quantum theory provides

Table 1.2 *A strategy exploiting the detection loophole for the CHSH test. In each round, Alice and Bob choose one of the eight quadruples of pre-established values that would give s = +2 (c.f., Table 1.1). Alice answers always, while Bob declines to answer to the input indicated by brackets, in such a way that the problematic output (boldface) is never produced.*

$a_0, a_1; b_0, b_1$	$a_0 b_0$	$a_0 b_1$	$a_1 b_0$	$a_1 b_1$
$+1, +1;\ +1, [+1]$	$+1$	N	$+1$	**N**
$-1, -1;\ -1, [-1]$	$+1$	N	$+1$	**N**
$+1, +1;\ +1, [-1]$	$+1$	**N**	$+1$	N
$-1, -1;\ -1, [+1]$	$+1$	**N**	$+1$	N
$+1, -1;\ [+1], +1$	N	$+1$	**N**	-1
$-1, +1;\ [-1], -1$	N	$+1$	**N**	-1
$+1, -1;\ [-1], +1$	**N**	$+1$	N	-1
$-1, +1;\ [+1], -1$	**N**	$+1$	N	-1

a very accurate description of such an experiment. It's very hard to believe that this accuracy is just an accident due to detectors being inefficient. Besides, the mechanism of the loophole assumes that a detector's firing depends on the input of the Bell test chosen in each round (to continue with the example, the polarization basis chosen for the measurement). But experimentalists know that the detector's firing depends on internal parameters: With respect to the inputs of the Bell test, the detector is performing a *fair sampling*.

This is why, rather than claiming failure, all the early experiments have reported the observation of nonlocality under the fair-sampling assumption. Most likely, several future experiments will legitimately continue to do so. That being said, it's also important to put to record that nonlocality without the fair sampling assumption has been observed, first in an experiment with entangled ions (Rowe *et al.*, 2001), then in several other platforms, including of course the three "loophole-free" experiments of 2015 cited previously.

Other loopholes related to the process of detection have been identified for some specific implementations: We refer the interested reader to the comprehensive review by Larsson (2014). Ultimately, they can all be closed by a rigorous implementation of the rule "one output for every input."

In summary, although they may prove challenging for some platforms, *the detection loopholes can be closed* and are therefore not a threat for the certification of nonlocality.

1.5.3 The "free-will loophole," or measurement independence

Instead of allowing the players to decline answering, the verifier could reveal some information about the inputs he is going to send out in every round. In the CHSH test, it is enough to inform the players about a pair of inputs that won't be used in a given round (in fact, more than enough: See Exercise 1.4).

This possibility sounds as artificial as the previous one, because the verifier has no reason to reveal that information. But there is a crucial difference. In the case of the detection loophole, it is easy to enforce an answer in every round (if the players refuse to comply, the verifier can fill the answer himself and too bad for them). Here, the verifier has to *ensure that no information leaks out to the players*. As frequent reports of leakage and hacking confirm, this is notoriously much more difficult to check, and ultimately impossible to guarantee in an absolute way.

In the laboratory setting, this loophole is open if the preparation of the system to be measured is correlated with the measurements that are going to be performed. This possibility is called *measurement dependence*; though less frequently than hacking, it has also been in the news.[10] The ultimate form of measurement dependence is super-determinism; leaving metaphysics for section 1.6, let us assume that measurement independence is possible in principle, and ask: How can one try and enforce it?

---

[10] In 2015 it was discovered that a car manufacturer had programmed some models to detect the specific procedures of an anti-pollution test. The car's emissions would then be lowered in order to pass that test, before returning to the normal, environment-unfriendly settings. As we see, measurement dependence exists; but it is rarely the result of an accident and is usually taken as the evidence of conscious tampering.

For some, the ultimate enforcement of measurement independence would be to choose the inputs using human free will; whence the name of "free-will loophole." I prefer to leave such a delicate notion as free will out of the picture.[11] At any rate, all that is required is to choose the inputs with *a process that is very unlikely to be correlated* with the resources shared by the players.[12] Some have gone as far as to generate the inputs of a Bell test from fluctuations of radiation coming from very distant stellar objects, i.e., produced by matter that has not been in contact with earthly matter since inflation, if ever (Handsteiner *et al.*, 2017). Impressive for a physicist, this choice of process may not be convincing for a technological skeptic: Who guarantees that those telescopes and electronics are really producing stellar randomness? This is why others have argued that the best way to approach the free will loophole is to generate the inputs from the letters of one's favorite book or the Geneva phonebook (Pironio, 2015). Surely there are several ways in which these inputs may not be called random: There is a structure in the text, and the information has been available in the Universe for quite some time—still, in order to refuse the evidence of Bell nonlocality, our skeptics must now believe that the behavior of some physical systems is correlated to the text of a book. If someone does not find this insane, there is little chance that they can be convinced anyway.

In summary: Whether viewed as leakage of information from a verifier, or as hidden correlations among the devices in a laboratory, we are in the presence of a loophole that *can only be closed under reasonable assumptions*. In this book, *we shall always assume that measurement independence holds* till section 11.4, where we shall see that one can relax it partially and still be able to certify Bell nonlocality. Finally, for interpretational matters it is crucial to stress that *measurement independence is compatible with determinism*: It requires that there exist several uncorrelated chains of events, but each chain can be deterministic. In other words, by assuming measurement independence, we are not introducing indeterminism by *fiat*, just as we are not required to accept any modality of human free will in order to certify nonlocality.

1.5.4   The "locality loophole," or hidden communication channels

The last loophole has a different flavor than the previous ones. It does not aim at generating a false positive, but a trivial positive: If the nonlocal resource could be *communication*, one says that the "locality loophole" is open. Closing the loophole would mean to design a Bell test, in such a way as to certify nonlocality while guaranteeing that no communication was happening. This seems to be possible using a physical fact: Information propagates at a speed bounded by that of light in vacuum.

---

[11] I cannot resist referring the reader to an introductory book that reviews the debate on free will, accessible to beginners like us physicists (Griffith, 2013).

[12] If the process that is used to choose the inputs is pseudo-random, it is a finite-length algorithm and ultimately the players would be able to learn it (Bendersky *et al.*, 2016). I cite this in a footnote because, important as it is conceptually, this possibility makes no difference in practice: Real Bell tests achieve excellent statistical significance in far fewer rounds than would be needed to guess even a moderately short algorithm.

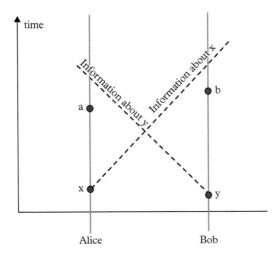

Figure 1.2 *The space-time configuration in which the locality loophole is closed, with the dotted lines representing the light cones. Information about x may arrive at the location of Bob only after the output b has been produced, and information about y may arrive at the location of Alice only after the output a has been produced. There is no absolute way of determining when the information about the inputs was created, nor when a definite output is produced: The events that are relevant to close the locality loophole can only be defined under reasonable assumptions.*

Recall that the information that needs to be communicated are the inputs. For every player p, we denote by $\vec{x}^{(p)}$ their position, by $t_i^{(p)}$ and $t_o^{(p)}$ the times at which the input is received, respectively at which the output is produced. The verifier wants to enforce that each player produce their output before any information about the other players' inputs may have arrived at their location: In jargon, *space-like separation* between the event "output" of each player and the events "input" of all the other players. This requirement reads[13] $|\vec{x}^{(p)} - \vec{x}^{(p')}| > c(t_o^{(p)} - t_i^{(p')})$ for all pairs of players (p, p') (Figure 1.2).

Can one find flaws in this argument? Some may question the physical assumption: Maybe some information can propagate faster than light. We shall discuss this option as interpretation in subsection 1.6.1, then as attempted models in chapter 11. But there is a more subtle possible flaw that went unnoticed for decades: In order to run an argument based on space-like separation, one must also be able to *identify the relevant events*. Alice can say when and where the input was fed into her devices; but she can't say when and where the information about the input was created in the Universe. Similarly, Alice can estimate when an electric current left the detector to convey the information to the computer; but is *that* the time at which the actual result is created, the time of the "collapse"?

---

[13] For notational simplicity, we assume that the relative positions of the players are fixed, which is usually the case in implementations. Also notice that, since the condition of space-like separation is Lorentz-invariant, we did not need to specify in which frame events are parametrized.

In summary, *the locality loophole can be closed for known communication channels* between the players, or by adopting operational definitions of the events, as several experiments did.[14] Ruling out any unknown form of communication seems to be impossible: Even if the speed of communication is believed to be bounded, we wouldn't know which are the relevant events.

1.5.5 The unknown loophole: Skepticism

It should be clear that Bell nonlocality has been scrutinized with great rigor. Some die-hard skeptics are not convinced: Every now and then, someone claims to have found the flaw in the argument. These claims are usually based on very convoluted arguments and tend to fall into three categories: Utterly wrong (for instance, the alleged counterexample is just a variation of the detection loophole); exegeses of Bell's papers (whether Einstein, Bell, or anyone else was right or wrong is interesting for the history of science, but science should be judged without reference to their authority); deep discussions on the meaning of probability (some of which may hit home but would apply to every statistical statement and not only to the certification of Bell nonlocality: After all, the philosophy of probability and the cogency of statistical conclusions are still debated).

To be sure, by definition one cannot exhaust the list of possible loopholes, and science should always be open to revision. But at this stage, I strongly believe that the burden of the proof should be on the deniers. If anyone has found the flaw, they should be able to write the corresponding algorithm, take two computers that have been pre-programmed together but do not communicate during the rounds of the game, and exhibit a statistically significant violation of a Bell inequality [see e.g., section 9 of (Gill, 2014)]. In the presence of such evidence, all physicists of the "establishment" will be ready to reconsider the matter.

1.6 Experimental Metaphysics?

There is abundant observational evidence for nonlocality with all the loophole closed – the memory and detection loopholes, indisputably; the free-will and locality loopholes, up to assumptions that go unquestioned in virtually all the rest of science, and have been noticed in this context only because of the strength of the claim. So, we are in the presence of a phenomenon that calls for interpretation.

We may want to start by refocusing on the dilemma. It is the same dilemma for any form of nonlocality, but let us just refer to the GHZ test: Given the outputs of two of

---

[14] The definition of the events is related to electric signals: For the input, when a signal leaves the "random number generator" to reach the measurement device; for the output, when the electric signal leaves the detector to propagate to the computer where the information will be stored. The locality loophole was first addressed in (Aspect, Dalibard, and Roger, 1982*a*), but the first experiment using a more proper random number generator was performed years later in the group of Anton Zeilinger (Weihs *et al.*, 1998). A few months earlier, the group of Nicolas Gisin had put the emphasis on the distance rather than on the timing of the random number generation, observing Bell nonlocality between players separated by 10km (Tittel *et al.*, 1998).

the players, there is only one possible output for the third—and nevertheless, that output was not predetermined. When was it determined then, and how?

Many positions have been put forward, often overlapping with one's interpretation of quantum theory. For the purpose of this book, I have chosen to classify the options in four groups. This being my own classification, to avoid both canonizations and imprecise attributions I have decided not to insert any citation in what follows. For further reading, one can start with some essays of very different, often opposite flavor published with the occasion of the 50th anniversary of Bell's theorem (Fuchs *et al.*, 2014; Maudlin, 2014; Werner, 2014; Wiseman, 2014; Żukowski and Brukner, 2014). There are also two systematic treatises on the meaning of Bell nonlocality, by Tim Maudlin (2011) and Jeffrey Bub (2015); and two popular books by Anton Zeilinger (2010) and Nicolas Gisin (2014)—and of course, a famous collection of John Bell's reflections (Bell, 2004).

1.6.1   Group 1: Nonlocal hidden variables

The easiest position to describe is that of those who infer from Bell nonlocality the existence of *nonlocal hidden variables*. With this position, one recovers determinism by staying within a *mechanical paradigm*: That is, one can simulate of how nature processes information in order to produce the outputs.

What would these nonlocal hidden variables be? A form of communication (superluminal or even retrocausal, i.e., propagating to the past), the infinitely rigid quantum ether of Bohmian mechanics, a connection in an unknown dimension ... Whatever they may be, in our $(3 + 1)$-dimensional space-time they would appear as "influences" carrying information from one location to the other, which Einstein famously dubbed "spooky action at a distance" in his debates with Bohr.[15]

Now, these hypothetical influences would carry information, only to tweak it in such a way that it looks no-signaling to us. With a positive wink, Shimony called it "peaceful coexistence with relativity". More negative critics rather highlight the conspiratorial flavor: Why would nature use a signal while hiding its use from us? Both this fine tuning and the relation will relativity will be discussed in detail in chapter 11. In particular, there we shall see a quantitative result: In order to remain "hidden," these influences must propagate at an *infinite* speed in their preferred frame.

1.6.2   Group 2: Superdeterminism and its friends

In the second group, I shall put superdeterminism and stances that (at least in my view) are akin to it.

In subsection 1.5.3, we have seen that Bell nonlocality can be demonstrated as soon as the processes that choose the inputs and those chosen by the players are independent. The strongest way to deny this measurement independence is *superdeterminism*: All the events in the Universe constitute a single, deterministic process. There could be somewhat milder ways of *denying measurement independence*: For instance, worldviews à

---

[15] The first record of this expression seems to date from the 1927 Solvay Conference.

la *The Matrix* in which our Universe is a big simulation, maybe not deterministic in origin (it could be run by aliens using their true free will). Needless to say, the consequences of adopting such a worldview extend far beyond solving the conundrum of Bell nonlocality.

Another option, directly inspired by quantum theory, are the so-called *many-worlds interpretations* that deny that a definite output is ever singled out. These interpretations say that the reversible dynamics of quantum theory is an accurate depiction of the deepest reality, which we do not perceive with our senses but have discovered with our investigations. What is usually deemed irreversible in an elementary reading of the quantum formalism, namely the act of measurement, is nothing else than getting entangled with the apparatus, and then with the environment, with the hard disk that stores the data, with the consciousness of the players ... The players will perceive definite outputs, and the verifier will certify Bell nonlocality after many rounds, because that's how the rules are set: In every world, there is indeterminacy and nonlocality. But nature is playing all the options, and this deployment of correlations is fully deterministic.

1.6.3   Group 3: Only statistics are speakable, a.k.a. the "orthodox"

Like the many-worlds interpretations, the third group also asserts the correctness of quantum formalism, but in a very different way. Here, quantum theory is the correct way of computing probabilities in the (only) physical world—and there is nothing else we should talk about. Individual rounds of a Bell test (or of any other experiment: interferometers, Stern-Gerlach ...) are "unspeakable."

As some of my colleagues like to say, this is "just standard quantum mechanics": Indeed, it is what people identify as the orthodox interpretation. But what should be mentioned, is that it implies a significant epistemological discipline. Physics is generally understood as a representation of nature, a study of the constituents of matter and their dynamics — loosely speaking, it should describe "how nature does it." However, a long list of philosophers may find this stance too naive, and the intrinsically statistical character of quantum theory has won several physicists over to the idea that *a law of nature may rather be a way of organizing our knowledge.* Bayesianism becomes the proper language: Probabilities capture someone's degree of belief. A law of physics does not prescribe the belief itself, which is subjective, but how beliefs should evolve given new information (and it is perfectly acceptable that these updating rules be not subjective).

If someone adopts this approach to physics and knowledge, the "intrinsic indeterminacy" of quantum theory adds only a minor element of discomfort: At some point, agents have to give up the possibility of a more refined description, one leading to stronger beliefs. As for the description of Bell tests, contrary to the many-worlds interpretation, the definite outputs are real and the agent has a special role.

The strength and weakness of this position are both simultaneously evident in the way it deals with the GHZ test: One refuses to explain how the third player manages to give that unique answer, but stresses than an agent should definitely bet on that unique answer, if informed about the other two. This position makes pragmatically correct statements and avoids all the problems. For some it's wisdom, for others escapism.

1.6.4 Group 4: Hoping for collapse

This last position, in a sense, closes the circle. In this view, the outputs are real (contrary to some in Group 2), they are produced through some intrinsically random process (contrary to Groups 1 and 2) and through Bell nonlocality we are learning about this process that does happen in nature (contrary to Group 3).

The challenge here is to describe the process, usually called "collapse," by which the output of each round is generated. There have been several attempts, but none seems to be fully convincing. Because of Bell nonlocality, any collapse model will have to be nonlocal to describe bipartite or multipartite statistics: The process that generates one player's output must take into account other players' inputs. But it seems inescapable that collapse models exhibit some form of nonlocality for single-player processes too: In a measurement of position, the particle must localize itself somewhere, whence the possibility of finding it elsewhere should fall to zero; if a single photon is sent through a beam-splitter and found in one of the beams, it must become impossible to find it in the other beam.

1.6.5 Additional remarks

I have just tried a simple systematization of the current state of interpretations. This matter is complex and can be approached from several angles. I'll go through a few more viewpoints here.

Let us first address the question of whether Bell nonlocality is *experimental metaphysics* that shapes our *Weltanschauung*. Group 1 would certainly claim so. For Groups 2 and 3, Bell nonlocality is just one of the rules that have been set up and does not play a foundational role (when it comes to Group 2, very few facts can claim to play a foundational role in a deterministic worldview). For Group 4, collapse models have first been studied to explain the appearance of a classical macroscopic world, but Bell nonlocality is something that such models are urged to explain too. At any rate, whatever position one reaches after reflection, Bell nonlocality must have entered that reflection: Nobody brushes it off as irrelevant *a priori*.

Next, I find it interesting to compare these interpretations in terms of *resources and information*. We don't have a recipe to describe how the outputs are generated if we stick only to resources that we can control. Group 4 hopes that this is still possible, at least with a special recipe of collapse. Group 1 favors a mechanistic recipe, at the price of introducing unobserved resources, the nonlocal hidden variables. Groups 2 and 3 stick to the resources that we have: For Group 2, all the outputs are generated according to the rules and it's sheer chance that we end up perceiving one rather than another alternative; for Group 3, how nature generates the outputs is not our business as long as our predictions are correct.

Further, let us consider the issue of *whether quantum theory is complete*, which was the title of the EPR paper. Both Groups 2 and 3 would definitely answer that quantum theory is complete; but we have noticed that they differ in what they call "quantum theory":

For the ones it's the kinematics and reversible dynamics, for the others the recipe for computing probabilities. Representatives of Group 4 would like to complete the theory with a model of collapse, probably hinting that collapse has always been a desired feature of quantum theory, a statement with which the others would vehemently disagree. Group 1 advocates for the need of quantum theory to be completed, at least in its ontology (our predictive power may well remain the same).

Philosophical labels are also worth mentioning. Based on observation, I can certify that most of my colleagues would like to be called "realist" in the philosophical sense of believing in the existence and intelligibility of an external reality. Conversely, when the debates get heated, it is frequent to hear insults like "idealist," "pragmatist," or "solipsist" thrown to the representatives of the other camp. One may wonder, for instance, where is the realism in Group 3: They would answer that the laws for updating our beliefs come to us from observation. All in all, this kind of labels should be avoided, also because few of us physicists would be able to pass an exam of philosophy on their exact meaning— philosophers themselves may not agree!

Another favorite topic of speculation is *where great figures of the past would stand in this debate.* The usual names that come up are of course Einstein, Bohr, and Bell himself. Assuming that he would stick to his aversion to a dice-playing God and to action-at-a-distance, Einstein would either think that we don't have yet enough evidence (something like Group 4) or lean towards determinism at higher levels (Group 2). For Bohr, it is clear that he would despise Group 1; if I'd have to bet, his sympathy would go to Group 3. We know more for Bell. He set out to construct a local hidden variable model, probably thinking that it was possible. When he found it is not, he shifted towards Bohmian mechanics, and when collapse models started to be studied he clearly looked at them with great hope. Where he would stand today, given the evidence that collapse models have not delivered much, is only guesswork.

Surely there is much more to it, and the readers will find further inspiration in reading more or in their own reflections. For the purpose of this book, it is time to put an end to this general introduction and to move on to the formalization of Bell nonlocality.

· ·

EXERCISES

Exercise 1.1 *Prove that the Mermin inequality* (1.10) *holds indeed for deterministic local variables. Hint: Either $c_0 = c_1$, or $c_0 = -c_1$.*

Exercise 1.2 *Re-derive Hardy's local variables (LV) prediction* (1.14) *by ticking out from Table 1.1 the quadruples of pre-established values that do not comply with the constraints (1.11)–(1.13).*

Exercise 1.3 *We consider a modification of the detection loophole strategy for CHSH described in Table 1.2. At every round, Alice and Bob choose one of the eight quadruples listed* *in the Table with probability $\frac{1}{8}$. Then Bob applies that strategy of declining with probability $1 - p$, whereas with probability p he produces the agreed output to whatever input he receives. The verifier computes S using only the rounds in which both Alice and Bob replied.*

Prove that the verifier will observe $S = \frac{4}{1+p}$ (hint: What is the fraction of rounds in which Bob replies?). Deduce that this strategy gives a false positive for every $p > 0$.

Exercise 1.4 *We consider a false positive for the CHSH test based on the free will loophole.*

1. *In every round, the verifier informs the players that one specific pair of inputs will be sent out with probability q, while the three other pairs will be equally probable. Find the value of S that the players can achieve with local strategies for every $q \in \left[0, \frac{1}{4}\right]$. Deduce that this leads to a false positive for every $q < \frac{1}{4}$, and that the quantum maximum $S = 2\sqrt{2}$ can be reached without having to set $p = 0$.*

2. *Consider now a different situation: The verifier informs the players that in every round the pair $(x, y) = (1, 1)$ is drawn with probability q and the other three pairs with equal probability. Does this open any loophole? Hint: The probabilities that enter a nonlocality test are conditional on the inputs.*

2

Formalizing Bell Nonlocality

L'universo [...] é scritto in linguaggio matematico e le lettere sono triangoli, cerchi e altre figure geometriche.

The Universe is written in mathematical language and its letters are triangles, circles and other geometrical figures.

Galileo, *Il saggiatore*

This chapter introduces the mathematical framework for the study of Bell nonlocality.

2.1 Bell Scenarios, Processes, and Behaviors

Virtually all the theory of Bell nonlocality is a study of *probability distributions*. This means that one is describing Bell tests in the asymptotic limit of infinitely many rounds. The application of this theory to real implementations, with their necessarily finite number of rounds, does not differ essentially from any laboratory use of statistical tools (see some specific remarks in section 2.6).

2.1.1 Bell scenarios

A *Bell scenario* is defined by specifying the number of players, the alphabet of inputs[1] that each player can receive, and the alphabet of possible outputs (the latter may be different for each input, but we shall only consider Bell tests in which the alphabet of outputs is the same for each input).

From now until chapter 5, we consider two-player, or bipartite, scenarios. The players are usually called Alice and Bob. The $(M_A, m_A; M_B, m_B)$ Bell scenario[2] is one in which Alice (Bob) has M_A (M_B) possible inputs and m_A (m_B) possible outputs for every input. In general statements, I shall adopt the notation $x \in \mathcal{X} = \{1, \ldots, M_A\}$, $a \in \mathcal{A} = \{1, \ldots, m_A\}$, $y \in \mathcal{Y} = \{1, \ldots, M_B\}$, $b \in \mathcal{B} = \{1, \ldots, m_B\}$. In concrete examples, it may be

---

[1] In the literature the inputs are often called "settings," since in the laboratory they determine which measurement to perform, i.e., how to set the measurement apparatus.

[2] For unknown reasons, the conventional usage in the literature is rather (M_A, M_B, m_A, m_B). I find it more logical to divide according to player, rather than specifying first the alphabet of the inputs and then that of the outputs.

Bell Nonlocality. Valerio Scarani. © Valerio Scarani 2019. Published in 2019 by Oxford University Press.
DOI: 10.1093/oso/9780198788416.001.0001

more convenient to adopt other conventions, for instance $\{0, 1\}$ or $\{-1, +1\}$ for binary alphabets, and I shall freely do it when no ambiguity is possible.

2.1.2 Processes and strategies

For every round, before receiving the inputs, the players agree on the *process* by which they will produce the outputs. At this point, the physical implementation of the process can be virtually anything: Computing a function, casting dice, and tossing coins, using quantum computers, discussing over the phone, or even operating as yet-unknown physical resources. The players obviously have to know what they are doing—but we don't need to: We treat the process as a black box and describe it only in terms of how the outputs are related to the inputs. This relation may be stochastic. Therefore, we identify a process λ with the collection of the probabilities $P_\lambda(a, b|x, y)$ of obtaining the outputs (a, b) when the inputs (x, y) were given:

$$\mathcal{P}_\lambda = \{P_\lambda(a, b|x, y) | \, a \in \mathcal{A}, b \in \mathcal{B}, x \in \mathcal{X}, y \in \mathcal{Y}\}. \tag{2.1}$$

The probabilities must be non-negative, i.e.,

$$P_\lambda(a, b|x, y) \geq 0 \ \ \forall a \in \mathcal{A}, b \in \mathcal{B}, \, x \in \mathcal{X}, y \in \mathcal{Y}, \tag{2.2}$$

and properly normalized for each pair (x, y), i.e.,

$$\sum_{a \in \mathcal{A}, b \in \mathcal{B}} P_\lambda(a, b|x, y) = 1 \ \ \forall x \in \mathcal{X}, \, y \in \mathcal{Y}. \tag{2.3}$$

For each pair of inputs (x, y), one has to give all the $P_\lambda(a, b|x, y)$ but one, because of the normalization. The number of real parameters needed to describe a process is therefore $D_{\text{gen}} = M_A M_B (m_A m_B - 1)$.

A *deterministic process* is one in which the outputs are uniquely determined by the inputs, i.e., $P_\lambda(a, \ b|x, \ y) = \delta_{(a, \, b) = F(x, \, y, \, \lambda)} = \delta_{a = f(x, \, y, \, \lambda)} \delta_{b = g(x, \, y, \, \lambda)}$. The number of deterministic processes is $\sharp_D = (m_A m_B)^{M_A M_B}$: For each pair of inputs, one has to say which pair of outputs is being produced.

We also have to say something about the players' *strategy* over all the rounds of the test:

- *Measurement independence is assumed throughout the book till chapter 11*: Thus, the choice of the process should not depend on the inputs of the round itself. In formal terms, λ and (x, y) are independent.

- We usually work with *Independent and identically distributed (i.i.d. strategies*. In an i.i.d. strategy, the process for each round is selected independently of what happened in previous rounds, by sampling from a given probability distribution $Q(\lambda)$, with $Q(\lambda) \geq 0$, and $\int d\lambda Q(\lambda) = 1$. As mentioned in subsection 1.5.1, the i.i.d. assumption should be taken critically: This will be addressed in section 2.6.

2.1.3   The observed behavior

The verifier does not know which process is used in each round. He observed coarse-grained statistics that are called the observed behavior, or simply *the behavior*, of the players:

$$\mathcal{P} = \{P(a,b|x,y)\,|\,a \in \mathcal{A}, b \in \mathcal{B}, x \in \mathcal{X}, y \in \mathcal{Y}\}, \qquad (2.4)$$

where

$$P(a,b|x,y) = \int d\lambda\, Q(\lambda) P_\lambda(a,b|x,y). \qquad (2.5)$$

By construction, the $P(a,b|x,y)$ are also non-negative and normalized:

$$P(a,b|x,y) \geq 0 \;\; \forall a \in \mathcal{A}, b \in \mathcal{B}, x \in \mathcal{X}, y \in \mathcal{Y} \qquad (2.6)$$

$$\sum_{a \in \mathcal{A}, b \in \mathcal{B}} P(a,b|x,y) = 1 \;\; \forall x \in \mathcal{X}, y \in \mathcal{Y}. \qquad (2.7)$$

Also, a generic behavior is specified by giving D_{gen} real parameters. The behavior is going to be *deterministic* if and only if the players use a given deterministic process λ_d in every round, i.e., $Q(\lambda) = \delta(\lambda - \lambda_d)$.

2.2   No-Signaling Processes and Behaviors

2.2.1   Definition and motivation

We introduce the following:

Definition 2.1 *A process λ is called no-signaling if each player's output statistics do not depend on other player's inputs, that is if*

$$P_\lambda(a|x,y) \overset{NS}{=} P_\lambda(a|x,y') \equiv P_\lambda(a|x) \text{ for all } a \in \mathcal{A}, x \in \mathcal{X}, y, y' \in \mathcal{Y}$$
$$P_\lambda(b|x,y) \overset{NS}{=} P_\lambda(b|x',y) \equiv P_\lambda(b|y) \text{ for all } b \in \mathcal{B}, y \in \mathcal{Y}, x, x' \in \mathcal{X}. \qquad (2.8)$$

Similarly, a behavior is called no-signaling if

$$P(a|x,y) \overset{NS}{=} P(a|x,y') \equiv P(a|x) \text{ for all } a \in \mathcal{A}, x \in \mathcal{X}, y, y' \in \mathcal{Y}$$
$$P(b|x,y) \overset{NS}{=} P(b|x',y) \equiv P(b|y) \text{ for all } b \in \mathcal{B}, y \in \mathcal{Y}, x, x' \in \mathcal{X}. \qquad (2.9)$$

These conditions are called the no-signaling constraints. The generalization to multipartite scenarios is straightforward.

Obviously, if all the $\mathcal{P}_\lambda$ satisfy the no-signaling condition, so will the behavior $\mathcal{P}$, but the converse is not true (Exercise 2.1).

When encountered for the first time, this important definition may sound convoluted: Wouldn't it be simpler to say that each player's output does not depend on the other players' input? Here lies exactly the subtle distinction between no-signaling and local processes and behaviors. The latter will be formally introduced in the next section, but the reader should recall the qualitative discussion of these issues in subsection 1.2.3.

2.2.2   Description of no-signaling statistics

Because of the no-signaling constraints, it takes fewer parameters than D_{gen} to describe a no-signaling process or behavior. The counting can be done as follows. One can first take the marginals as independent parameters: There are $M_A(m_A - 1)$ independent $P(a|x)$, and $M_B(m_B - 1)$ independent $P(b|y)$. Consider now any choice of (x, y): For every fixed $b = \beta$, once the marginal $P(a)$ is given, one is left with $m_A - 1$ independent numbers $P(a, \beta)$; similarly, for every fixed $a = \alpha$, once the marginal $P(b)$ is given, one is left with $m_B - 1$ independent numbers $P(\alpha, b)$. All in all, a no-signaling process can be fully specified by giving

$$D_{\text{NS}} = M_A(m_A - 1)M_B(m_B - 1) + M_A(m_A - 1) + M_B(m_B - 1) \qquad (2.10)$$

independent real numbers.

In line with this parameter counting, we can introduce the convenient *Collins-Gisin representation* of a no-signaling process or behavior (Collins and Gisin, 2004). We show it for the Bell scenarios $M_A = M_B = 2$ and $m_A = m_B = 2, 3$:

$$\mathcal{P} \overset{NS}{=} \begin{array}{c|c|c}
 & P(a=1|x=1) & P(a=1|x=2) \\
\hline
P(b=1|y=1) & P(1,1|1,1) & P(1,1|2,1) \\
\hline
P(b=1|y=2) & P(1,1|1,2) & P(1,1|2,2)
\end{array}, \qquad (2.11)$$

$$\mathcal{P} \overset{NS}{=} \begin{array}{c|cc|cc}
 & P(a=1|x=1) & P(a=2|x=1) & P(a=1|x=2) & P(a=2|x=2) \\
\hline
P(b=1|y=1) & P(1,1|1,1) & P(2,1|1,1) & P(1,1|2,1) & P(2,1|2,1) \\
P(b=2|y=1) & P(1,2|1,1) & P(2,2|1,1) & P(1,2|2,1) & P(2,2|2,1) \\
\hline
P(b=1|y=2) & P(1,1|1,2) & P(2,1|1,2) & P(1,1|2,2) & P(2,1|2,2) \\
P(b=2|y=2) & P(1,2|1,2) & P(2,2|1,2) & P(1,2|2,2) & P(2,2|2,2)
\end{array}.$$

The number of entries of the table are indeed $D_{\text{NS}} = 8, 24$. All the probabilities that are not written explicitly can be reconstructed from the given ones: For instance, in the second case, one has $P(a = 3|x) = 1 - P(a = 1|x) - P(a = 2|x)$; $P(3, b|x, y) = P(b|y) - P(1, b|x, y) - P(2, b|x, y)$; etc. The example is easy to generalize to any other Bell scenario: There must be entries for each value of the inputs, while one element of each output alphabet can be skipped since its statistics can be reconstructed from the rest; and the marginals can be taken out precisely because of the no-signaling constraints, that make them independent of the other player's input.

2.3   Local Behaviors

2.3.1   Definition and motivation

In chapter 1 we introduced the idea of locality in a Bell test: *In each round*, the process λ must be such that each player's output is generated taking only the same player's input into account. At the level of probability distributions, this translates into the following definition, which is obviously central to this book:

Definition 2.2 *A process is called local if it is of the form*

$$P_\lambda(a,b|x,y) \overset{LV}{=} P_\lambda(a|x)P_\lambda(b|y), \tag{2.12}$$

where we have used the traditional label "Local Variables."

   A behavior is called local if it can be written as a convex combination of local processes, as in (1.1). *A behavior that cannot be written that way is called Bell-nonlocal, or simply nonlocal.*

Local processes and behaviors clearly satisfy the no-signaling constraints, thus they can also be described by D_{NS} real numbers and cast in the Collins-Gisin representation. But a local process is much more constrained, *ad nauseam*, let's us compare the two definitions in words:

- Local process: In each round, each player's output is independent of the other players' inputs.
- No-signaling process: The marginal distribution of each player's output is independent of the other players' inputs.

2.3.2   Local deterministic processes

Local processes don't need to be deterministic, but the class of local deterministic processes plays an important role.

Definition 2.3 *A process $\mathcal{P}_{LD,\lambda}$ is a local deterministic (LD) process if, for any input, the output is deterministic:*

$$P_{LD,\lambda}(a|x) = \delta_{a=f(x,\lambda)}, P_{LD,\lambda}(b|y) = \delta_{b=g(y,\lambda)}. \tag{2.13}$$

An equivalent way of characterizing a LD process consists in just giving the list of outputs for all possible inputs:

$$\lambda_{LD} \equiv \{a_1, a_2, \ldots, a_{M_A}; b_1, b_2, \ldots, b_{M_B}\} \in \mathcal{A}^{|\mathcal{X}|} \times \mathcal{B}^{|\mathcal{Y}|}, \qquad (2.14)$$

the link with the previous notation being $a_x = f(x, \lambda)$ and $b_y = g(y, \lambda)$. From this notation, it is obvious that the number of LD processes is

$$\sharp_{LD} = m_A^{M_A} m_B^{M_B}. \qquad (2.15)$$

These LD processes are all and the only deterministic processes that satisfy the no-signaling constraints (2.8); all the remaining $\sharp_D - \sharp_{LD}$ deterministic processes can only be realized with signaling resources (Exercise 2.2).

2.3.3 Characterization of local behaviors: Fine's theorem

By definition of Bell locality, the behavior seen by the verifier is a *local behavior*, if and only if there exist a family of local processes λ and a distribution $Q(\lambda)$ such that

$$P(a, b|x, y) \overset{LV}{=} \int d\lambda\, Q(\lambda)\, P_\lambda(a|x) P_\lambda(b|y). \qquad (2.16)$$

This expression, which we have already encountered (1.1), is the main mathematical object in the formalization of Bell nonlocality: We'll need to find ways to prove that a given behavior is *not* of this form. As it turns out, this general form is not very useful because the number of processes λ and the distribution $Q(\lambda)$ are arbitrary. Fortunately, there exist two compact characterizations of local behaviors:

Theorem 2.1 *These two equivalent results are both due to Arthur Fine (1982):*

  (a) A behavior $\mathcal{P}$ is local if and only if it is a convex mixture of local deterministic processes:

$$P(a, b|x, y) \overset{LV}{=} \sum_{j=1}^{m_A^{M_A}} \sum_{k=1}^{m_B^{M_B}} q_{jk}\, \delta_{a=f_j(x)} \delta_{b=g_k(y)} \qquad (2.17)$$

  with $\lambda \equiv (j, k)$ and $\sum_{j,k} q_{jk} = 1$.

  (b) Using the notation (2.14): A behavior $\mathcal{P}$ is local if and only if there exists a joint probability distribution $\mathbf{P}(a_1, \ldots a_{M_A}; b_1, \ldots, b_{M_B})$, such that each $P(a, b|x, y)$ can be obtained as the marginal

$$P(a, b|x, y) \overset{LV}{=} \sum_{\{a_j | j \neq x\}} \sum_{\{b_k | k \neq y\}} \mathbf{P}(a_1, \ldots a_{M_A}; b_1, \ldots, b_{M_B}). \qquad (2.18)$$

Proof Let us start from statement (a). The "if" direction is obvious, because (2.17) is a special case of (2.16). For the reverse, given $P(a, b|x, y)$ of the form (2.16), one can compute the cumulative distribution $\Sigma(a) = \sum_{\alpha \leq a} P_\lambda(\alpha|x)$, where we use the labeling $a \in \mathcal{A} = \{1, \ldots, m_A\}$ of the outputs (labelings being arbitrary, one can redo the proof for any other choice, at the expense of a more cumbersome notation). Let us then add a new local parameter $\mu_A \in [0, 1]$ and declare that a is chosen according to the following deterministic rule:

$$P_{\lambda, \mu_A}(a|x) = \begin{cases} 1 & \text{if } \Sigma(a-1) \leq \mu_A < \Sigma(a), \\ 0 & \text{otherwise.} \end{cases} \tag{2.19}$$

If μ_A is drawn with uniform distribution, the original process is recovered:

$$\int_0^1 d\mu_A \, P_{\lambda, \mu_A}(a|x) = \int_{\Sigma(a-1)}^{\Sigma(a)} d\mu_A = P_\lambda(a|x).$$

The same construction can be made on Bob's side. Therefore (2.16) can be rewritten as

$$P(a, b|x, y) \overset{LV}{=} \int d\lambda Q(\lambda) \int_0^1 d\mu_A \int_0^1 d\mu_B P_{\lambda, \mu_A}(a|x) P_{\lambda, \mu_B}(b|y).$$

Now, as we said, each $P_{\lambda, \mu_A}(a|x)$ is deterministic, so it must be equal to one of the $\delta_{a=f_j(x)}$; and similarly, each of the $P_{\lambda, \mu_B}(b|y)$ must be equal to one of the $\delta_{b=g_k(y)}$. The weights q_{jk} are just the integrals of the measure $Q(\lambda)d\lambda d\mu_A d\mu_B$ over the corresponding sets. This concludes the proof of (a). From here, the proof of statement (b) is straightforward. Indeed, on the one hand, LD processes fix the value of the output for all possible inputs, so $(a) \Longrightarrow (b)$ for

$$\mathbf{P}\big(a_1 = f_j(1), \ldots, a_{M_A} = f_j(M_A); b_1 = g_k(1), \ldots, b_{M_B} = g_k(M_B)\big) \equiv q_{jk}.$$

On the other hand, any assignment of values to all the a_x, respectively b_y, defines a valid $f_j(x)$, respectively $g_k(y)$; so $(b) \Longrightarrow (a)$.

2.3.4 Examples of local behaviors

It is useful to look at a few examples of local behaviors.

- The behavior of *one player* can always be reproduced with LV and thus with a mixture of deterministic strategies. In the case of quantum behaviors, λ may contain the description of the quantum state ρ: The player can then compute on paper the expected probability distribution for any measurement, then generate the outputs according to that distribution. The fresh reader will find this statement obvious, and such indeed it is. More sophisticated readers may have problems with notion accumulated in their formation: Did not great teachers like Feynman and Schwinger derive the quantum formalism from the single-spin Stern-Gerlach measurement? Isn't it established that there cannot be a LV model for states with negative Wigner function? These concerns are addressed in Appendix D.

- One of the most obvious bipartite behaviors that can be described with LV is exactly the one that is presented in popular lore as an astonishing feat of quantum entanglement: Two players *produce always the same output when queried with the same input*. Even without mathematics, it is clear that this is so, because we can achieve the same in everyday life.[3] For instance, we could ask two persons to dress identically, then send them to different location: If queried about the same piece of their apparel, they will provide the same answer. What is astonishing with entanglement, is that this happens while the outputs are not pre-established; and we know that they are not, because some other measurements can be used to prove Bell nonlocality.

- Every bipartite behavior in which *one of the players has only one input* is also local. Indeed, if $|\mathcal{X}| = 1$, Alice's *actual* input in each round is known in advance, so it's known by Bob too. In every round, the process λ will determine Alice's output and distribute Bob's outputs accordingly for each of his possible inputs.

- More subtle than the previous, let's suppose that Bob's outputs $\underline{b} = (b_1, \ldots, b_M)$ have a joint probability distribution $P(\underline{b})$ independent of Alice's inputs.[4] For every input x of Alice, it follows from the previous point that one can construct $P(a_x, \underline{b})$ such that summing over all Bob's inputs but the desired b_y yields $P(a_x, b_y)$. Moreover, the independence of Bob's distribution from Alice's inputs requires $\sum_{a_x} P(a_x, \underline{b}) = P(\underline{b})$ for all x. Then, we can construct the joint probability distribution

$$P(a_1, .., a_M, \underline{b}) = \frac{\prod_{x=1}^{M_A} P(a_x, \underline{b})}{P(\underline{b})^{M_A - 1}} \tag{2.20}$$

such that all marginals match the behavior. Indeed, by first summing over all Alice's inputs but the desired a_x, one recovers $P(a_x, \underline{b})$; at which point, by construction the sum over all but one of Bob's inputs yields $P(a_x, b_y)$.

- If each player can produce perfect copies of the information shared to define the process, the resulting behavior will be local. Indeed, each player can produce as many copies as his/her inputs, then apply the procedure to obtain the corresponding output, before the run of the test. This is a situation of pre-determined outputs. Thus, all behaviors that can be obtained with perfectly copiable information can also be obtained with pre-determined outputs (for a small modification, see Exercise 2.3).

---

[3] This is a variation over the criterion set forward in one of the earliest reviews on Bell nonlocality (Werner and Wolf, 2001*b*): If it can be done with ping-pong balls, it's not interesting.

[4] Notice that this is stronger than requesting that the bipartite behavior is no-signaling: It's requesting no-signaling even if Bob could actually obtain the outputs corresponding to all his inputs in each round. In quantum theory, which is formally no-signaling, this would always be the case; an obvious case in which $P(\underline{b})$ exists is that of Bob using a set of commuting measurements.

2.4 The Local Polytope and Bell Inequalities

In this section, we study the set $\mathcal{L}$ of all local behaviors, in short the *local set*. We introduce a geometric characterization which is fundamental for the study of Bell nonlocality. The famous Bell inequalities are the conditions for a behavior $\mathcal{P}$ to belong to $\mathcal{L}$.

2.4.1 The local set as a polytope

The first simple property of $\mathcal{L}$ is *convexity* (Figure 2.1, left): For all $p \in [0, 1]$ and for all choice of $\mathcal{P}_{LV,1} \in \mathcal{L}$ and $\mathcal{P}_{LV,2} \in \mathcal{L}$, the convex sum $\mathcal{P} = p\mathcal{P}_{LV,1} + (1-p)\mathcal{P}_{LV,2}$ belongs to $\mathcal{L}$ too. The proof is straightforward: If we use the representation (2.17) for $\mathcal{P}_{LV,1}$ and $\mathcal{P}_{LV,2}$, the convex sum $\mathcal{P}$ is also of the same form with $q_{jk} = pq_{jk,1} + (1-p)q_{jk,2}$. Convexity is also an intuitive necessity: One possible way of realizing $\mathcal{P}$ is to imagine that in a fraction p of the rounds one is sampling from $\mathcal{P}_{LV,1}$, and in the other rounds one is sampling from $\mathcal{P}_{LV,2}$. Clearly, which behavior is used in each round can be decided in advance.

Because probabilities are bounded, the set $\mathcal{L}$ is compact. Now, a compact convex set is the convex hull of its extremal points,[5] i.e., those points that cannot be written as a convex sum of other points in the set. *The extremal points of $\mathcal{L}$ are all the local deterministic behaviors and only those.*

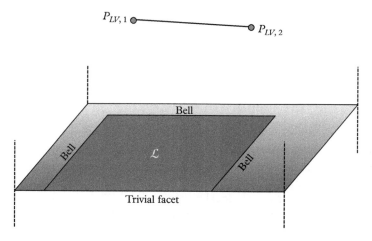

Figure 2.1 *Top: Geometry of the convexity of the local set: If two local behaviors are represented by points in a probability space, the whole line connecting them is also made of local behaviors. Bottom: Sketch of the local polytope $\mathcal{L}$ as a subset of the set of no-signaling behaviors (light grey), both embedded in $\mathbb{R}^{D_{NS}}$, with $D_{NS} = 2$ for the sake of the illustration (the minimal Bell scenario has $D_{NS} = 8$). The local polytope as drawn has four extremal points, one trivial facet and three tight Bell inequalities (see subsection 2.4.3). The dotted lines in the third dimension remind that the no-signaling set is a set of zero measure in the space of all possible behaviors.*

---

[5] This is very intuitive; the formal proof goes under the name of Krein-Milman theorem.

Let us prove this statement: We know from Fine's theorem 2.1 that any $\mathcal{P} \in \mathcal{L}$ can be written as a convex sum of LD behaviors: Thus, only LD behaviors may be extremal. *Ex absurdo*, suppose that a LD behavior can be decomposed as convex combination $\mathcal{P}_{LD,?} = p\mathcal{P}_{LD,1} + (1 - p)\mathcal{P}_{LD,2}$ with $0 < p < 1$ and $\mathcal{P}_{LD,1} \neq \mathcal{P}_{LD,2}$. The latter two behaviors must differ for at least one pair of inputs (x, y): For these inputs, there are $(\alpha_1, \beta_1) \neq (\alpha_2, \beta_2)$ such that $P_{LD,1}(\alpha_1, \beta_1|x, y) = 1$, and $P_{LD,2}(\alpha_2, \beta_2|x, y) = 1$. But then $P_{LD,?}(\alpha_1, \beta_1|x, y) = p$ and $P_{LD,?}(\alpha_2, \beta_2|x, y) = 1 - p$: So $\mathcal{P}_{LD,?}$ is not deterministic, contradicting the assumption that it was LD.

Thus, the local set is a bounded polyhedron or *polytope* embedded in $\mathbb{R}^{D_{NS}}$ (Figure 2.1, right). Since $\sharp_{LD} > D_{NS}$, some extremal points are linearly dependent. The $(D_{NS} - 1)$-dimensional hyperplanes that delimit it are called *facets* and are in finite number. To be a facet, a hyperplane must satisfy two properties. First, to ensure the dimensionality, at least D_{NS} linearly independent extremal points must lie on the hyperplane (unless the zero vector is chosen as to belongs to the facet, in which case the number is $D_{NS} - 1$). Second, all the extremal points that do not lie on the hyperplane must be found on the same side of it. Mathematically, let $v \in \mathbb{R}^{D_{NS}}$ is the vector normal to the facet and oriented outside the polytope: If the equation for the points $\mathcal{P}$ of the facet is $v \cdot \mathcal{P} = f$, then

$$v \cdot \mathcal{P} \leq f \text{ for all } \mathcal{P} \in \mathcal{L}. \tag{2.21}$$

It is clear that a polytope is fully determined by listing its extremal points, because this information is sufficient to find all the facets. Alternatively, a polytope is also fully determined by listing its facets: Their one-dimensional intersections are the extremal points. In the case of $\mathcal{L}$, the extremal points are easy to list because they are just the LD behaviors, while finding the facets requires some computation.[6]

2.4.2   Assessing the locality of a behavior

Assessing *whether a given behavior* $\mathcal{P}$ *is local* is the elementary task of Bell nonlocality. Essential in theoretical studies, this assessment is also part of the more complex task of assessing whether a real, finite set of data is compatible with local realism, which will be discussed in section 2.6. It is an instance of a "membership problem," since it is asking whether the behavior belongs to the local polytope $\mathcal{L}$.

We just noticed that it's easy to write down the extremal points $\mathcal{P}_{LD,i}$, so we are going to formulate the problem as: Find $\underline{q} = \{q_1, \ldots, q_{\sharp_{LD}}\}$ such that

$$q_i \geq 0 \ \forall i, \qquad \sum_{i=1}^{\sharp_{LD}} q_i = 1, \qquad \mathcal{P} = \sum_{i=1}^{\sharp_{LD}} q_i \mathcal{P}_{LD,i}. \tag{2.22}$$

---

[6] We shall see in section 9.1 that the opposite is the case for another polytope relevant for nonlocality, the so-called no-signaling polytope: There, the facets are easy to list, and one has to work to find the extremal points.

This is a *feasibility linear program*, one the simplest instances of convex optimization. The standard reference for convex optimization is the book of Boyd and Vandenberghe (2004), and we provide a very elementary introduction as Appendix E. Here it is enough to know that, while it is usually impossible to give a solution in closed form even for a linear program, algorithms are known, and packages implementing them exist for most mathematical softwares. These algorithms generate the so-called *dual program* of the original one (which is then called the *primal*) and search for solutions for both.

For the problem under study, the duality of linear programs reflects the dual characterization of the polytope in terms of extremal points or in terms of facets. Indeed, as we show in Appendix E.2.1, the dual of the program (2.22) is a program that checks whether there is a facet that separates the behavior $\mathcal{P}$ from the whole local polytope. The algorithm has therefore two possible outputs: If it finds a solution $\underline{q}$ for the primal, it returns this solution, proving that $\mathcal{P} \in \mathcal{L}$; otherwise, it returns the equation of a facet that separates $\mathcal{P}$ from $\mathcal{L}$, proving that $\mathcal{P} \notin \mathcal{L}$. In either case, the verification of the correctness of the output is straightforward.

Although it is obvious from what precedes, I want to highlight explicitly that *the locality of behaviors can be assessed without any prior knowledge of the facets*—that is, in the terminology that we are going to introduce, of the Bell inequalities. This is going to be important, since listing all the facets is a daunting task for most Bell scenarios, as we shall see in chapters 4 and 5.

2.4.3   The notion of Bell inequality

A criterion that separates *some* nonlocal behavior from *all* LV behaviors is generically called *Bell inequality*. Usually, one considers *linear* Bell inequalities, i.e., criteria of the from

$$I(\mathcal{P}) \equiv \sum_{a,b,x,y} v_{abxy} P(a,b|x,y) \leq I_L. \tag{2.23}$$

As mentioned, the most natural candidates are the *facets of the local polytope*. The number of facets being finite, it is in principle possible to list them all for any Bell scenario.

Not all the facets of $\mathcal{L}$ are candidates for Bell inequalities: All the $M_A M_B m_A m_B$ conditions $P(a, b|x, y) \geq 0$ define facets,[7] called *positivity facets* or *trivial facets*.

The equations (2.23) of the non-trivial facets of the local polytope are called *tight Bell inequalities*. Non-tight Bell inequalities are suboptimal for the mere certification of

---

[7] This is not completely trivial: Since $\mathcal{L}$ satisfies also no-signaling, one should check if some of the positivity conditions happen to coincide when the no-signaling constraints (2.9) are enforced. As it turns out, this may happen, but only in uninteresting scenarios: Notably, if $m_B = 1$ i.e., the only output for Bob is 0, then the no-signaling condition on the marginal of Alice reads $P(a|x) = P(a, 0|x, y)$ for all y; in other words, all the conditions $P(a, 0|x, y) \geq 0$ are the same. Also, one may wonder why there are no facets associated to the other boundary, $P(a, b|x, y) \leq 1$. The reason is that the normalization $\sum_{a,b} P(a,b|x,y) = 1$ is always assumed, and consequently the upper bound on each $P(a, b|x, y)$ is a consequence of the positivity of all the other probabilities.

nonlocality but may have additional interesting features, we shall encounter examples in subsection 4.2.3 and in section 5.3.

2.4.4   Multiple versions, liftings, and no-signaling forms of an inequality

Every Bell inequality appears in *multiple versions* in probability space, due to the freedom of *relabeling*: If a criterion tests nonlocality, the same criterion in which some of the labels have been changed will also test nonlocality in the same way. The possible relabelings are the following:

- One can relabel the inputs of each player. Specifically, if π is a permutation over M_A elements and π' is a permutation over M_B elements, then

$$v_{abxy} \longrightarrow v_{ab\pi(x)\pi'(y)} \tag{2.24}$$

  defines another version of the same inequality.

- Once the labeling of the inputs is chosen, for each input one is free to relabel the outputs. Specifically, for each x we write π_x a permutation over m_A elements, and for each y we write π'_y a permutation over m_B elements: Then

$$v_{abxy} \longrightarrow v_{\pi_x(a)\pi'_y(b)xy} \tag{2.25}$$

  defines another version of the same inequality.

- The most commonly studied Bell scenarios are such that $(M_A, m_A) = (M_B, m_b)$. In this case, the multiplicity can be further increased by a factor 2 by permuting the name of the players, i.e.,

$$v_{abxy} \longrightarrow v_{bayx} \ [\text{if} (M_A, m_A) = (M_B, m_b)]. \tag{2.26}$$

This counting shows that there can be as many as $M_A! M_B! (m_A!)^{M_A} (m_B!)^{M_B}$ versions of a given inequality, or even twice as many if the players can be exchanged. However, several relabelings may lead to the same version, so the multiplicity of an inequality is usually smaller. The multiplicity must be taken into account: For $\mathcal{P}$ to be local, one must check that $\mathcal{I}(\mathcal{P}) \leq \mathcal{I}_L$ holds for *all versions* of all the Bell inequalities in the scenario. This may be bothersome on paper, but recall that a simple linear program does it without the need to give any Bell inequality a priori.

Besides, when moving from a Bell scenario to a larger one (more players and/or more inputs and/or more outputs), all the inequalities of the smaller scenarios are inherited in several versions called *liftings*. As an example, suppose that $\{v_{abxy}|a, b, x, y \in \{0,1\}\}$ defines a Bell inequality in the $(2,2; 2,2)$ scenario: If we move to the $(3,2;2,2)$ Bell scenario, we can already anticipate three Bell inequalities, defined by $\{v_{abxy}|x \in \{0,1\}, a,b,y \in \{0,1\}\}$, $\{v_{abxy}|x \in \{0,2\}, a,b,y \in \{0,1\}\}$, and $\{v_{abxy}|x \in \{1,2\}, a,b,y \in \{0,1\}\}$—each of course with its multiplicity due to relabelings. If the original inequality is a facet of the local polytope, the liftings will also be facets of the local polytope for the larger scenario.

Finally, we must mention that, given a version of an inequality, one can transform the v_{abxy} almost beyond recognition by exploiting the normalization of probabilities and even more the no-signaling constraints (2.9). Just to give an idea, consider $a \in \{0,1\}$: With normalization, one can write $1 = P(a = 0|x,y) + P(a = 1|x,y)$, and using no-signaling one can even write $1 = P(a = 0|x,y) + P(a = 1|x,y')$ with $y' \neq y$. We shall refer to different *forms* of an inequality, and we shall see several examples, starting from the next section.

Ultimately, all these are symmetries that can be elegantly and efficiently tackled with tools of group theory (Rosset, Bancal, and Gisin, 2014). In this book, we shall deal with these issues in a rather pedestrian way, limiting ourselves to specific examples and occasional warnings.

2.5 The CHSH Scenario

The simplest Bell scenario is $(2,2;2,2)$, usually known as *CHSH scenario* from the work of (Clauser *et al.*, 1969) that we have already encountered in section 1.3.

2.5.1 The local polytope and the CHSH inequality

In this scenario, the local polytope is embedded in $\mathbb{R}^8$ and has $\sharp_{LD} = 16$ extremal points, each being the product of one deterministic process for Alice $[(a_0, a_1) = (0,0), (0,1), (1,0), \text{or } (1,1)]$ with one for Bob [same lists for (b_0, b_1)]. We have already listed these points, with a different labeling, in Table 1.1. In the Collins-Gisin representation, the extremal points read

$$
\mathcal{P}_{(0,0)(0,0)} = \begin{array}{c|c|c} & 1 & 1 \\ \hline 1 & 1 & 1 \\ \hline 1 & 1 & 1 \end{array}, \mathcal{P}_{(0,0)(0,1)} = \begin{array}{c|c|c} & 1 & 1 \\ \hline 1 & 1 & 1 \\ \hline 0 & 0 & 0 \end{array}, \ldots, \mathcal{P}_{(1,1)(1,1)} = \begin{array}{c|c|c} & 0 & 0 \\ \hline 0 & 0 & 0 \\ \hline 0 & 0 & 0 \end{array}
$$

$$(2.27)$$

where the completion of the list is left as Exercise 2.4. Having this list, the local polytope is fully determined, and one can set out to find the facets. Even in this simplest of cases, the analytical derivation is tedious, we give it in Appendix G.1 for a meaningful sub-polytope. The complete polytope has 24 facets (Froissart, 1981; Fine, 1982; Collins and Gisin, 2004): The expected 16 positivity facets and 8 facets defined by

$$I^{(1)} = E_{00} + E_{01} + E_{10} - E_{11} \le 2$$
$$I^{(2)} = E_{00} + E_{01} - E_{10} + E_{11} \le 2$$
$$I^{(3)} = E_{00} - E_{01} + E_{10} + E_{11} \le 2$$
$$I^{(4)} = -E_{00} + E_{01} + E_{10} + E_{11} \le 2$$
$$I^{(5)} = -E_{00} + E_{01} + E_{10} + E_{11} \ge -2$$
$$I^{(6)} = E_{00} - E_{01} + E_{10} + E_{11} \ge -2$$
$$I^{(7)} = E_{00} + E_{01} - E_{10} + E_{11} \ge -2$$
$$I^{(8)} = E_{00} + E_{01} + E_{10} - E_{11} \ge -2$$

where

$$E_{xy} = P(a = b|x,y) - P(a \ne b|x,y). \tag{2.28}$$

It is manifest that these are just versions of the same inequality up to relabelings: $I^{(2)}$ is obtained from $I^{(1)}$ by permuting the labels of Bob's inputs, $I^{(3)}$ is obtained from $I^{(2)}$ by permuting the roles of Alice and Bob (or by permuting the labels of both Alice's and Bob's inputs), etc. Due to the symmetry of the expression, we have only 8 instead of the potential $2M_A!M_B!(m_A!)^{M_A}(m_B!)^{M_B} = 128$ versions. Thus, up to relabelings, *there is only one tight inequality in the CHSH scenario, namely the CHSH inequality*

$$S = E_{00} + E_{01} + E_{10} - E_{11} \le 2. \tag{2.29}$$

This is the same CHSH expression (1.3) that we derived in section 1.3. There, we took advantage of the smart labeling $a, b \in \{-1, +1\}$ to write $\langle a_x b_y \rangle = \sum_{a,b} ab P(a,b|x,y) = E_{xy}$. The current derivation, with E_{xy} defined by (2.28), highlights that any labeling of the outputs is allowed, as indeed it should be.

2.5.2   The CH form of the CHSH inequality

Among the infinitely many forms of the CHSH inequality equivalent under no-signaling, the most frequently encountered one is

$$S_{CH} = \sum_{x,y=0}^{1} (-1)^{xy} P(0,0|x,y) - P(a = 0|x = 0) - P(b = 0|y = 0) \le 0, \tag{2.30}$$

called *CH form* because it was introduced by Clauser and Horne (1974). To prove the equivalence, one can plug into (2.29) the expression

$$E_{xy} = 4P(0,0|x,y) - 2P(a=0|x) - 2P(b=0|y) + 1, \qquad (2.31)$$

that follows from the relations

$$P(0,1|x,y) = P(a=0|x) - P(0,0|x,y),$$
$$P(1,0|x,y) = P(b=0|y) - P(0,0|x,y),$$
$$P(1,1|x,y) = 1 - P(0,0|x,y) - P(0,1|x,y) - P(1,0|x,y)$$
$$= 1 - P(a=0|x) - P(b=0|y) + P(0,0|x,y).$$

The CH expression contains exactly the terms that appear in the Collins-Gisin representation. If the behavior $\mathcal{P}$ is written in that representation, one can conveniently write the Bell expression as a term-by-term multiplication of tables:

$$S_{CH} = \mathcal{I}_{CH} \cdot \mathcal{P}, \quad \text{with} \quad \mathcal{I}_{CH} = \begin{array}{c||c|c|} & -1 & 0 \\ \hline\hline -1 & 1 & 1 \\ \hline 0 & 1 & -1 \\ \hline \end{array}. \qquad (2.32)$$

By the further replacements $P(a=0|x=0) - P(0,0|0,1) = P(0,1|0,1)$ and $P(b=0|y=0) - P(0,0|1,0) = P(1,0|1,0)$, S_{CH} can be rewritten in the *Eberhard form* (Eberhard, 1993)

$$S_{Ebe} = P(0,0|0,0) - P(0,1|0,1) - P(1,0|1,0) - P(0,0|1,1) \le 0. \qquad (2.33)$$

The numerical values of the three expressions are related through

$$S_{CH} = S_{Ebe} = \frac{1}{4}S - \frac{1}{2}. \qquad (2.34)$$

A historical remark must be made at this point. Clauser and Horne presented (2.30) as a new inequality, which is perfectly understandable for those early days. Unfortunately, before the notions of no-signaling and of the local polytope were fully formalized, the belief that CHSH and CH are inequivalent was reinforced by a misunderstanding in the approach to the detection loophole;[8] as received knowledge, it has persisted in the literature till very recently.

---

[8] The first studies of the detection loophole addressed it correctly with the CH form, but somehow failed to see how to address it with the CHSH form. As a result, the conviction spread that the latter is simply unsuited for the task. In Appendix B.3 we show how to approach the detection loophole both with the CHSH and the CH form, and obtain the same result.

2.6   Proper Statistics: Beyond i.i.d. and Finite Samples

In this last section, we prove that the local polytope remains unchanged if the underlying strategy is not i.i.d., then discuss some subtleties that one should take into account for a proper statistical assessment of a Bell test.

2.6.1   Non-i.i.d. behaviors and the local polytope

In this chapter, we have started from a definition of behaviors (2.5) that assumes an i.i.d. strategy. We have built the local polytope from the corresponding definition of i.i.d. local behaviors. The i.i.d. assumption must be removed: As mentioned in subsection 1.5.1, if the players could pass a Bell test by just adopting a local non-i.i.d. strategy, Bell nonlocality wouldn't be worth much discussion.[9]

For behaviors that describe an asymptotically large number of rounds, we are going to prove that the definition of the local polytope does not change if one relaxes the i.i.d. assumption. For simplicity of notation, we consider processes λ that correlate *two* consecutive rounds, but the proof extends easily to any number of rounds. If two rounds are known to be correlated, the proper description of the behavior is[10] $P(a_1, a_2, b_1, b_2 | x_1, x_2, y_1, y_2)$. Given this, the single-round statistics can be computed through

$$P'(a,b,x,y) = \frac{1}{2} \left[\sum_{a_2,b_2,x_2,y_2} P(a,a_2,b,b_2|x,x_2,y,y_2)P(x,x_2,y,y_2) \right. $$
$$\left. + \sum_{a_1,b_1,x_1,y_1} P(a_1,a,b_1,b|x_1,x,y_1,y)P(x_1,x,y_1,y) \right]$$

where the factor $\frac{1}{2}$ comes from the normalization $\sum_{a,b,x,y} P'(a,b,x,y) = 1$. Thus, the effective single-round behavior is

$$P'(a,b|x,y) = P'(a,b,x,y)/P'(x,y), \tag{2.35}$$

where $P'(x,y) = \frac{1}{2}[\sum_{x_2,y_2} P(x,x_2,y,y_2) + \sum_{x_1,y_1} P(x_1 x, y_1, y)]$. Now, suppose that the two-round behavior $P(a_1, a_2, b_1, b_2 | x_1, x_2, y_1, y_2)$ is *local*: By Fine's theorem, it is a convex mixture of local deterministic two-round behaviors. But, any LD two-round behavior is the product of two LD single-round behaviors (the fact that the two rounds may not be independent simply defines the rule with which each single-round behavior is chosen). Thus $P'(a, b|x, y)$ is a convex mixture of LD single-round behaviors, which

---

[9] Still, it took almost forty years after Bell's paper for this matter to be properly addressed by Richard Gill [as early as 2001, see references in (Gill, 2014)] and (Barrett *et al.*, 2002).

[10] It should be clear that, in this subsection, a_j does not mean $a_{x=j}$, but the output of Alice in the j-th round.

means it's local. *A contrario*, if $P'(a, b|x, y)$ were nonlocal, $P(a_1, a_2, b_1, b_2|x_1, x_2, y_1, y_2)$ must have been nonlocal too.

Notice that the proof does not requires the rounds to be sequential: The result holds even in the case of parallel queries, where the verifier sends all the inputs to the players and elicit all the outputs. In fact, the critical assumption is measurement independence, that we assume systematically till the last chapter (and we'll see in subsection 11.4.3 that, in the presence of measurement dependence, a local two-round behavior may lead to a nonlocal effective single-round behavior).

2.6.2   Certifying nonlocality on real data

Asymptotic statements like the previous one provide a simple working ground for theorists. But real data always come in finite amount, so one has to discuss fluctuations. Among pioneering work, one must cite that of Richard Gill, synthetized in (Gill, 2014), and the first dismissal of the memory loophole by (Barrett *et al.*, 2002). Several improvements were reported around the time of the loophole-free Bell tests, we refer to (Zhang *et al.*, 2013; Elkouss and Wehner 2016) for a first introduction. These tools are still being refined at the moment of writing: For instance, a Bayesian approach under the i.i.d. assumption has been provided in (Gu *et al.*, 2019).

For the level of this book, it is worth pointing out one peculiarity: In the case of nonlocality, *unprocessed raw data fail to provide a point estimator*. The reason is that the local polytope is by construction a subset of the set of no-signaling behaviors, but the

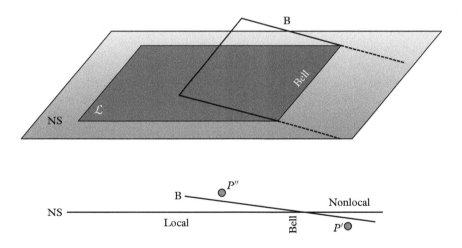

Figure 2.2 *If observed frequencies are taken as estimators of probabilities, the estimated behavior does not satisfy the no-signaling constraint rigorously. Bell inequalities are defined only in the no-signaling set (NS), because so is the local polytope (L): Thus, plugging frequencies into one's favorite form of a Bell inequality is, strictly speaking, ambiguous. This is illustrated in the figure: The estimated signaling behavior P' (respectively, P'') is manifestly a fluctuation of a nonlocal (respectively, local) behavior. However, if the inequality is written as B, P' does not violate it while P'' does.*

estimator of the behavior obtained from frequencies over a finite sample will deviate from the strict no-signaling condition. Plugging such an estimator into the linear program (2.22) will always return a negative result. It's even worse if one plugs the signaling estimator into their favorite form of a Bell inequality[11] (subsection 2.4.4): Given a signaling behavior and a Bell inequality as facet of the local polytope, one can always find both a form of the inequality that is violated by the behavior and a form that is not (Figure 2.2). If one wants a point estimator, one first has to project the estimator obtained from the frequencies into the no-signaling set (Lin *et al.*, 2018). The error estimator for this procedure is not known at the moment of writing.

....

EXERCISES

Exercise 2.1 *Construct an example of a no-signaling behavior as a coarse-graining of signaling processes.*

Exercise 2.2 *Prove that, in any Bell scenario, deterministic behaviors are either local deterministic or signaling.*

Exercise 2.3 *Alice and Bob share some resource prior to the Bell test. Suppose that Alice's information can be perfectly copied (while we make no such assumption on Bob's). Prove that the resulting behavior is local.*

Exercise 2.4

    (a) *Complete the list (2.27) by writing down all sixteen LD behaviors in Collins-Gisin representation.*

    (b) *For each, compute the value of the CH form (2.30). Compare it with the CHSH value given in Table 1.1 and provide a geometric interpretation of this observation.*

---

[11] Unfortunately a large number of experimental papers have done exactly that over the years. I do not doubt that the data could have supported nonlocality (maybe up to loopholes), but the claim is not strictly substantiated.

3

Bell Nonlocality in Quantum Theory

> *To avoid entanglements and interferences had long been one of her first principles.*
>
> C.S. Lewis, *That hideous strength*

While Bell nonlocality is defined and formalized without reference to quantum theory, it is this theory that describes accurately the known *no-signaling nonlocal resources*. This chapter introduces the study of nonlocality with the quantum formalism.

3.1 Quantum Behaviors

3.1.1 Quantum and nonlocality: A preamble

In subsection 1.2.3 we made the point that Bell nonlocality is intriguing because it can be demonstrated with some no-signaling resources—or more precisely, resources that are accurately described by a formalism that implies no-signaling: Quantum theory.

In order not to disrupt the flow of the study of nonlocality, I won't pause in the main text to introduce this theory. Many readers will likely be very familiar with it. Elementary knowledge at the undergraduate level is assumed, while the more advanced notions are summarily introduced in Appendix C with references for further study.

Quantum theory provides a further very important insight. Nonlocality is not associated to specific physical degrees of freedom (the charge of the electron, the hyperfine constant, the frequency of a light mode ...). What matters is the possibility of creating and measuring suitable states of composite systems, described as *entangled* by the theory. As long as one can manipulate those states, any physical degrees of freedom goes.

3.1.2 Definitions

In quantum theory, the shared resource is described as a *quantum state* shared among the players. A behavior $\mathcal{P}$ is obtained by each player performing a local measurement on their subsystem.

The joint system of Alice and Bob is described by the tensor product Hilbert space $\mathcal{H}_A \otimes \mathcal{H}_B$. A state ρ is described by a positive Hermitian operator of trace 1. A m_A-output measurement on Alice's system is described by $\mathcal{M}_A^x = \{\Pi_a^x | a \in \mathcal{A}\}$, where the Π_a^x are

Bell Nonlocality. Valerio Scarani. © Valerio Scarani 2019. Published in 2019 by Oxford University Press.
DOI: 10.1093/oso/9780198788416.001.0001

positive Hermitian operators on $\mathcal{H}_A$ satisfying $\sum_a \Pi_a^x = \mathbb{I}$. A m_B-output measurement on Bob's system is described by $\mathcal{M}_B^y = \{\Pi_b^y | b \in \mathcal{B}\}$, where the Π_b^y are analogously defined. Notice that the measurements do not need to be projective: They can be generalized measurements (a.k.a. positive-operator valued measures, POVMs).

A behavior $\mathcal{P}$ is called *quantum behavior* if there exist a state ρ and families of measurements $\{\mathcal{M}_A^x | x \in \mathcal{X}\}$ and $\{\mathcal{M}_B^y | y \in \mathcal{Y}\}$, such that

$$P(a,b|x,y) = \mathrm{Tr}(\rho \Pi_a^x \otimes \Pi_b^y). \tag{3.1}$$

In chapter 2, from the definition of local behavior (2.16) we derived a compact characterization of the set of all such behaviors, namely the local polytope. The discussion of the set of all quantum behaviors is more complex. Although it could fit here, I will postpone it to chapter 6: I prefer the reader to first become familiar with several *examples* of nonlocality in quantum theory.

3.1.3 Relations between states, measurements, and behaviors

Bell nonlocality is defined on behaviors, while quantum theory is phrased in terms of states and measurements. Here we spell out some evident but important consequences of this difference.

The *states* are the shared resources. In any Bell scenario, given a state, infinitely many behaviors are possible because of the freedom of choosing the measurements. We can introduce here the notion of *state-behavior* as the collection of the statistics of all possible local measurements on ρ:

$$\mathcal{P}_Q(\rho) = \left\{ P(a,b|\vec{a},\vec{b}) = \mathrm{Tr}(\rho \Pi_a(\vec{a}) \otimes \Pi_b(\vec{b})) \,\middle|\, a \in \mathcal{A}, b \in \mathcal{B}, \forall \vec{a}, \vec{b} \right\} \tag{3.2}$$

where $\vec{a}, \vec{b}$ parametrize the measurements. In principle, this parametrization should include POVMs, but often for simplicity one restricts state-behaviors to projective measurements (to practice this notion, see Exercise 3.1). States that are equivalent up to local unitaries produce the same state-behavior. Indeed, if $\tilde{\rho} = U\rho U^\dagger$ with $U = U_A \otimes U_B$, then $\mathrm{Tr}(\rho \Pi_a^x \otimes \Pi_b^y) = \mathrm{Tr}(\tilde{\rho} \tilde{\Pi}_a^x \otimes \tilde{\Pi}_b^y)$ where $\tilde{\Pi}_a^x = U_A \Pi_a^x U_A^\dagger$ and $\tilde{\Pi}_b^y = U_B \Pi_b^y U_B^\dagger$.

To prove that ρ is a nonlocal resource, it is sufficient to find the families of measurements that generate a nonlocal behavior. Conversely, to prove that ρ is not a nonlocal resource, one should prove that all the choices of measurements lead to behaviors that can be simulated with LVs. Thus, the assessment of nonlocality for a quantum resource ρ involves *a priori* a heuristic step: Finding the suitable measurements or constructing an LV model. That being said, the nonlocality of two classes of states has been clarified with generic arguments:

- It follows immediately from the expression (3.1) that *separable states* can generate only local behaviors. Indeed, inserting $\rho = \sum_k q_k \rho_A^k \otimes \rho_B^k$ into (3.1) one obtains $P(a,b|x,y) = \sum_k q_k \mathrm{Tr}(\rho_A^k \Pi_a^x) \mathrm{Tr}(\rho_B^k \Pi_b^y)$, which is of the form (2.16) for all measurements.

- We shall see in section 3.2.4 that *all pure entangled states* are nonlocal resources. The proof is constructive and provides a choice of measurements that generate a nonlocal behavior (without any claim of optimality, in any sense one may want to give to this word).

As for the nonlocality of *mixed entangled states*, we shall review the approaches and main results in section 3.3.

Let us now turn to the properties of the *measurements*. If a player uses a set of jointly-measurable[1] POVMs, it will be possible to construct the joint probability distribution of all their outputs. In this case, that player can't contribute to a nonlocal behavior as proved in subsection 2.3.4. Conversely, in order to create a nonlocal behavior, it is a necessary to use non-jointly-measurable POVMs. One may ask whether this is also sufficient, but the answer is negative. Indeed, there are explicit examples of non-jointly-measurable POVMs for one player, such that the resulting behavior will be local, irrespective of the number and choice of measurements of the other player (Hirsch *et al.*, 2018; Bene and Vértesi, 2018).

3.2   CHSH in Quantum Theory

3.2.1   The CHSH Bell operator

We shall use the notation $x, y \in \{0, 1\}$ for the inputs and $a, b \in \{-1, +1\}$ for the outputs, and work with the standard form (2.29) of the CHSH inequality. The correlation coefficients for given inputs (2.28) are now expressed as

$$E_{xy} = \mathrm{Tr}\left[\rho(\Pi_{+1}^x \otimes \Pi_{+1}^y + \Pi_{-1}^x \otimes \Pi_{-1}^y - \Pi_{+1}^x \otimes \Pi_{-1}^y - \Pi_{-1}^x \otimes \Pi_{+1}^y)\right]$$
$$= \mathrm{Tr}\left[\rho(\Pi_{+1}^x - \Pi_{-1}^x) \otimes (\Pi_{+1}^y - \Pi_{-1}^y)\right] \equiv \mathrm{Tr}\left[\rho A_x \otimes B_y\right], \tag{3.3}$$

where $A_x = \Pi_{+1}^x - \Pi_{-1}^x$ and $B_y = \Pi_{+1}^y - \Pi_{-1}^y$ are Hermitian operators of unspecified dimensions, whose eigenvalues lie between -1 and $+1$, and are equal to these values in the case of projective measurements. The value of the CHSH expression (2.29) in state ρ can be seen as the average value $S = \mathrm{Tr}(\rho \mathbf{S})$ of the *CHSH operator*

$$\mathbf{S} = A_0 \otimes B_0 + A_0 \otimes B_1 + A_1 \otimes B_0 - A_1 \otimes B_1. \tag{3.4}$$

In particular, *a quantum state ρ violates CHSH if and only if there exist measurements such that* $\mathrm{Tr}(\rho \mathbf{S}) > 2$.

Before continuing, we stress that a Bell test does *not* consist of single-shot measurements of a Bell operator. The Bell operator is self-adjoint, and as such it constitutes

---

[1] The textbook equivalence between joint measurability and commutation holds only for projective measurements. For POVMs, commutation implies joint measurability but the converse is not true: There exist POVMs whose elements do not commute but are nonetheless jointly measurable (more in Appendix C).

a valid quantum observable. But it cannot be measured by separated players because (as can be expected and we shall prove in the next subsection) some of its eigenstates may be entangled. The measurements of the players are still described by the POVM elements Π_a^x and Π_b^y. The operators A_x, the B_y and $\mathbf{S}$ are derived from those and used for mathematical convenience.

3.2.2   The Tsirelson bound

We proceed now to prove the famous *Tsirelson[2] bound* (Cirel'son, 1980):

$$S \leq 2\sqrt{2} \quad \text{for quantum behaviors.} \tag{3.5}$$

The bound is derived as an upper bound for the largest eigenvalue of $\mathbf{S}$ in absolute value, $\|\mathbf{S}\|_\infty = \max_\psi |\langle\psi \mathbf{S}\psi\rangle|$. A very direct proof follows from noticing[3] that $\mathbf{S}$ can be rewritten as

$$\mathbf{S} = \frac{1}{\sqrt{2}}(A_0^2 + A_1^2 + B_0^2 + B_1^2) - \frac{\sqrt{2}-1}{8}\text{[positive operator]}, \tag{3.6}$$

where the positive operator is given by

$$\left((\sqrt{2}+1)(A_0 - B_0) + A_1 - B_1\right)^2 + \left((\sqrt{2}+1)(A_0 - B_1) - A_1 - B_0\right)^2$$
$$+ \left((\sqrt{2}+1)(A_1 - B_0) + A_0 + B_1\right)^2 + \left((\sqrt{2}+1)(A_1 + B_1) - A_0 - B_0\right)^2.$$

Since the second term of the sum is non-positive, we have $\|\mathbf{S}\|_\infty \leq \|\frac{1}{\sqrt{2}}(A_0^2 + A_1^2 + B_0^2 + B_1^2)\|_\infty$. The bound (3.5) follows from $\|A_x\|_\infty \leq 1$, $\|B_y\|_\infty \leq 1$, through the triangle inequality $\|A + B\|_\infty \leq \|A\|_\infty + \|B\|_\infty$ and the fact that $\|\mathbf{A}^2\|_\infty = \|\mathbf{A}\|_\infty^2$ for normal operators.

   A more elegant proof can be provided assuming that the measurements are projective. In fact, no generality is lost in this assumption, because the dimensions d_A and d_B of the Hilbert spaces of Alice and Bob are left unspecified: By Naymark's theorem, any POVM in a given dimension can be implemented as a projective measurement in a higher dimension. For projective measurements, the eigenvalues of the A_x and B_y are exactly $+1$ and -1 (degenerate if $d_A, d_B > 2$); whence $\|A_x\|_\infty = \|B_y\|_\infty = 1, A_x^2 = \mathbb{I}_{d_A}$ and $B_y^2 = \mathbb{I}_{d_B}$. With these equalities, the *square* of the CHSH operator reads

$$\mathbf{S}^2 = 4\mathbb{I}_{d_A} \otimes \mathbb{I}_{d_B} - [A_0, A_1] \otimes [B_0, B_1]. \tag{3.7}$$

---

[2] The romanization of this name was changed from Cirel'son to Tsirelson in the 1980s.
[3] It takes a Tsirelson to notice this.

Besides,

$$||[A_0,A_1]||_\infty = ||A_0A_1 - A_1A_0||_\infty \leq ||A_0A_1||_\infty + ||A_1A_0||_\infty \leq 2||A_0||_\infty||A_1||_\infty = 2,$$

where for the last estimate we have used the inequality $||A_0A_1||_\infty \leq ||A_0||_\infty||A_1||_\infty$. Similarly one finds $||[B_0,B_1]||_\infty \leq 2$. The triangle inequality leads then to $||\mathbf{S}^2||_\infty \leq 8$, whence (3.5) follows.

Besides being more elegant, this second proof makes manifest that $[A_0, A_1] = 0$ or $[B_0, B_1] = 0$ will imply $S \leq 2$ (a special instance of the fact that joint measurability leads to local behaviors). It is also the starting point to prove that the state and the measurements that achieve (3.5) are essentially unique (this "self-testing" character will be presented in section 7.1).

3.2.3   CHSH for two-qubit states and projective measurements

The canonical example of violation of CHSH is that by the *maximally entangled state of two qubits*. We have already written such a state in (1.5). Up to local unitaries, it can be cast in the form of the singlet state

$$|\Psi^-\rangle = \frac{1}{\sqrt{2}}\left(|+\hat{n}\rangle|-\hat{n}\rangle - |-\hat{n}\rangle|+\hat{n}\rangle\right), \tag{3.8}$$

where the choice of the direction $\hat{n}$ that defines the bases does not matter because the state is invariant by bilateral rotation. Projective measurements on a qubit are labeled by a direction $\hat{m} \in \mathbb{S}^2$, the surface of the unit sphere embedded in $\mathbb{R}^3$. Specifically, $\mathcal{M}^{\hat{m}} = \{\Pi^{\hat{m}}_{+1}, \Pi^{\hat{m}}_{-1}\}$ with $\Pi^{\hat{m}}_s = \frac{1}{2}(\mathbb{I} + s\hat{m} \cdot \vec{\sigma})$ and where $\vec{\sigma}$ are the Pauli matrices.[4] A standard calculation (Exercise 3.2) shows that

$$P(a,b|\hat{a}, \hat{b}) = \langle \Psi^- |\Pi^{\hat{a}}_a \otimes \Pi^{\hat{b}}_b| \Psi^- \rangle = \frac{1}{4}(1 - ab\,\hat{a} \cdot \hat{b}). \tag{3.9}$$

This means that the correlation coefficient (2.28) is $E_{\hat{a},\hat{b}} = -\hat{a} \cdot \hat{b}$. In the context of a Bell test, for the sake of not carrying on a bothersome minus sign, there is no harm in assuming that Bob systematically flips his output, or that he uses a reference frame that is related to Alice's by parity inversion. So, starting from a singlet state, one can generate a behavior in which $E_{xy} = \hat{a}_x \cdot \hat{b}_y$, and the CHSH expression reads

$$S(\Psi^-) = \hat{a}_0 \cdot \left(\hat{b}_0 + \hat{b}_1\right) + \hat{a}_1 \cdot \left(\hat{b}_0 - \hat{b}_1\right). \tag{3.10}$$

---

[4] Having used x, y as inputs of Bell tests, I shall use the notation $\hat{x}$ etc., when a direction in space is meant. For instance, the Pauli matrix that is usually written σ_x will be written $\sigma_{\hat{x}}$.

Let's proceed to find the measurement directions that maximize this value. We note that the sum and difference of two unit vectors are orthogonal, so we can write $\hat{b}_0 + \hat{b}_1 = 2\cos\chi\,\hat{c}$ and $\hat{b}_0 - \hat{b}_1 = 2\sin\chi\,\hat{c}^\perp$. The parameter χ is defined by $\hat{b}_0 \cdot \hat{b}_1 = \cos 2\chi$. Given this observation, obviously the best choice to maximize $S(\Psi^-)$ is $\hat{a}_0 = \hat{c}$ and $\hat{a}_1 = \hat{c}^\perp$. With this choice one has $S(\Psi^-) = 2(\cos\chi + \sin\chi)$. Finally, we can maximize this expression over χ to find $S(\Psi^-) = 2\sqrt{2}$ for $\cos\chi = \sin\chi = \frac{1}{\sqrt{2}}$. So, the maximally entangled state of two qubits saturates the Tsirelson bound (3.5) by generating the behavior

$$P(a,b|x,y) = \frac{1}{4}\left(1 + ab(-1)^{xy}\frac{1}{\sqrt{2}}\right). \tag{3.11}$$

That gives the correlators $E_{xy} = (-1)^{xy}\frac{1}{\sqrt{2}}$. Working back through the optimization, we can find the measurement directions that must be chosen to reach that maximum. One of them, say $\hat{a}_0$, is arbitrary, since the state is invariant by rotation; then $\hat{a}_1$ is any direction orthogonal to $\hat{a}_0$. Having thus chosen Alice's measurement directions, Bob's are given by $\hat{b}_0 = \frac{1}{\sqrt{2}}(\hat{a}_0 + \hat{a}_1)$ and $\hat{b}_1 = \frac{1}{\sqrt{2}}(\hat{a}_0 - \hat{a}_1)$, so they are also orthogonal.

As it turns out, with not too much additional effort, one can derive the maximal value of the CHSH expression for *any two-qubit state, pure or mixed, under projective measurements* (Horodecki, Horodecki, and Horodecki, 1995). Indeed, any two-qubit state is of the form

$$\rho = \frac{1}{4}\left(\mathbb{I}\otimes\mathbb{I} + \vec{r}_\rho\cdot\vec{\sigma}\otimes\mathbb{I} + \mathbb{I}\otimes\vec{s}_\rho\cdot\vec{\sigma} + \sum_{i,j=\hat{x},\hat{y},\hat{z}} T_\rho^{ij}\sigma_i\otimes\sigma_j\right). \tag{3.12}$$

With this notation, $E_{xy} = \hat{a}_x \cdot T_\rho\hat{b}_y$ where T_ρ is the 3×3 matrix with entries T_ρ^{ij}; so the CHSH expression reads

$$S(\rho) = \hat{a}_0 \cdot T_\rho\left(\hat{b}_0 + \hat{b}_1\right) + \hat{a}_1 \cdot T_\rho\left(\hat{b}_0 - \hat{b}_1\right). \tag{3.13}$$

Since $||\hat{a}_0|| = 1$, the maximal value of $\hat{a}_0 \cdot T_\rho\hat{c}$ is $||T_\rho\hat{c}||$, obtained when $\hat{a}_0$ is chosen parallel to $T_\rho\hat{c}$. By the same reasoning on the term involving $\hat{a}_1$, we reach

$$\max_{\hat{a}_0,\hat{a}_1,\hat{b}_0,\hat{b}_1} S(\rho) = \max_{\hat{b}_0,\hat{b}_1} 2\left(\cos\chi\,||T_\rho\hat{c}|| + \sin\chi\,||T_\rho\hat{c}^\perp||\right)$$

$$= \max_{\hat{c},\hat{c}^\perp} 2\sqrt{||T_\rho\hat{c}||^2 + ||T_\rho\hat{c}^\perp||^2}$$

where in the last step we used the well-known optimization $\max_\chi x\cos\chi + y\sin\chi = \sqrt{x^2 + y^2}$ achievable with the choice $\cos\chi = x/\sqrt{x^2 + y^2}$. Finally, $||T_\rho\hat{c}||^2 = \hat{c}\cdot T_\rho^t T_\rho\hat{c}$

where T_ρ^t the transpose of T_ρ. Now, $T_\rho^t T_\rho$ is a positive symmetric matrix, therefore the maximization is achieved by choosing $\hat{c}$ and $\hat{c}^\perp$ as the two eigenvectors associated to its two largest eigenvalues λ_1 and λ_2:

$$\max_{\hat{a}_0, \hat{a}_1, \hat{b}_0, \hat{b}_1} S(\rho) = 2\sqrt{\lambda_1 + \lambda_2}. \tag{3.14}$$

It is instructive, and useful for what follows, to work out the case of *pure two-qubit states* more explicitly. By the Schmidt decomposition, any two-qubit pure state is equivalent up to local unitaries to

$$|\Psi(\theta)\rangle = \cos\theta \, |+\hat{z}\rangle |+\hat{z}\rangle + \sin\theta \, |-\hat{z}\rangle |-\hat{z}\rangle, \, \theta \in \left[0, \frac{\pi}{4}\right]. \tag{3.15}$$

Thus,

$$T_{\Psi(\theta)} = \begin{pmatrix} \sin 2\theta & 0 & 0 \\ 0 & -\sin 2\theta & 0 \\ 0 & 0 & 1 \end{pmatrix}. \tag{3.16}$$

The maximal achievable value of CHSH is

$$\max_{\hat{a}_0, \hat{a}_1, \hat{b}_0, \hat{b}_1} S(\Psi(\theta)) = 2\sqrt{1 + \sin^2 2\theta}, \tag{3.17}$$

which is always larger than 2, unless $\sin 2\theta = 0$. Therefore, *all pure entangled states of two qubits violate CHSH* for suitable measurements, and only maximally entangled ones saturate the Tsirelson bound. Let us determine the corresponding measurement settings. The eigenvector associated to the largest eigenvalue of $T_\rho^t T_\rho$ is $\hat{c} = \hat{z}$; the orthogonal subspace being degenerate, we can choose any vector in it as $\hat{c}^\perp$; in other words, there are many different choices of measurement directions leading to the maximal violation. Let us choose $\hat{c}^\perp = \hat{x}$. With this choice,

$$\hat{b}_{0,1} = \cos\chi\,\hat{z} \pm \sin\chi\,\hat{x} \text{ with } \cos\chi = 1/\sqrt{1 + \sin^2 2\theta}. \tag{3.18}$$

Furthermore, $\hat{a}_0$ must be the unit vector parallel to $T_\rho \hat{c}$, and $\hat{a}_1$ to $T_\rho \hat{c}^\perp$, so here

$$\hat{a}_0 = \hat{z}, \; \hat{a}_1 = \hat{x}. \tag{3.19}$$

In summary, Alice's measurement directions are orthogonal, $\hat{a}_0$ being the direction that defines the Schmidt basis ($\hat{z}$ in our convention); Bob's two measurement directions are not orthogonal, but are symmetrical around the same direction and lie in the same plane as Alice's.

3.2.4   All pure bipartite entangled states violate CHSH

As an extension of the previous, let us now prove that all *bipartite* pure entangled state (i.e., of any dimensionality) violates CHSH. Any d-dimensional bipartite pure state can be written in its Schmidt decomposition $|\Psi\rangle = \sum_{k=1}^{d} c_k |k\rangle |k\rangle$, $c_k \in \mathbb{R}^+$, where the bi-orthogonal basis is chosen such that $c_1 \geq c_2 \geq \ldots \geq c_d \geq 0$. The state is entangled if and only if $c_2 \neq 0$. Let us rewrite the state as

$$|\Psi\rangle = \sum_{j=1}^{\lceil d/2 \rceil} \sqrt{p_j} \left(\cos\theta_j |2j-1\rangle |2j-1\rangle + \sin\theta_j |2j\rangle |2j\rangle \right)$$

$$\equiv \sum_{j=1}^{\lceil d/2 \rceil} \sqrt{p_j} |\Psi_j(\theta_j)\rangle \tag{3.20}$$

where $p_j = c_{2j-1}^2 + c_{2j}^2$, $\cos\theta_j = \frac{c_{2j-1}}{\sqrt{p_j}}$ and $\sin\theta_j = \frac{c_{2j}}{\sqrt{p_j}}$ (if d is odd, $\sin\theta_{\lceil d/2 \rceil} \propto c_{d+1} = 0$). From the previous section, we know that CHSH can take the value[5]

$$S(\Psi) = \sum_{j=1}^{\lceil d/2 \rceil} p_j 2\sqrt{1 + \sin^2 2\theta_j} \tag{3.21}$$

with measurements of the form

$$A_x = \bigoplus_{j=1}^{\lceil d/2 \rceil} \hat{a}_{x,j} \cdot \vec{\sigma}_j, \quad B_y = \bigoplus_{j=1}^{\lceil d/2 \rceil} \hat{b}_{y,j} \cdot \vec{\sigma}_j \tag{3.22}$$

and the suitable choice of unit vectors in each subspace. Since $\sum_j p_j = 1$, $S(\Psi) > 2$ as soon as $c_2 > 0$.

All in all, we have proved that *all pure bipartite entangled states are nonlocal resources.* This is often referred to as "Gisin's theorem," since Nicolas Gisin was the first to ask the question and to answer it for bipartite states (Gisin, 1991) with a proof similar to the one we have just given. Shortly after, Popescu and Rohrlich (1992a), who were working independently on the same line, extended the proof to multipartite states (see section 5.1.2).

---

[5] For the current purposes, a lower bound is sufficient. The *maximal* violation of CHSH by an arbitrary pure state is not known analytically beyond the case of two qubits; however, there is strong numerical evidence that the lower bound used here is actually tight.

3.3 Mixed Entangled States as Nonlocal Resources

While all pure entangled states are nonlocal resources, the case of mixed entangled states is more complex. This is not surprising, since the same happens in entanglement theory. At a glance, the situation is as follows.

As first proved by Reinhard Werner (1989), some mixed entangled states do not exhibit any single-copy nonlocality: That is, any quantum behavior (3.1) obtainable by measuring them is local. The original example is presented in subsection 3.3.1; a few other families of such states have been found since then, but we don't have any general criterion [see (Augusiak *et al.*, 2014) for a review].

However, this is not the end of the story. If the players share a source that produces the entangled state ρ, which is local for single-copy measurements, they can try to create a nonlocal resource by many-copy processing. Subsection 3.3.2 reviews these protocols.

3.3.1 Single-copy nonlocality: Werner states

Consider the single-parameter family of two-qubit states ("Werner states") defined as

$$\rho_W = W \left|\Psi^-\right\rangle \left\langle\Psi^-\right| + (1 - W)\frac{\mathbb{I}}{4}, \ W \in \left[-\frac{1}{3}, 1\right]. \tag{3.23}$$

The state-behavior that describes the statistics of all possible projective measurements on ρ_W is

$$\mathcal{P}_Q(W) = \left\{ P(a, b|\hat{a}, \hat{b}) = \frac{1}{4}(1 - abW\hat{a} \cdot \hat{b}) \,\middle|\, a, b \in \{-1, +1\}, \hat{a}, \hat{b} \in \mathbb{S}^2 \right\}. \tag{3.24}$$

Using the criterion of negative partial transposition, it can be proved that Werner states are separable for $W \leq \frac{1}{3}$ and entangled otherwise (see Appendix C). Nonetheless, the behavior $\mathcal{P}_Q(W)$ can be reproduced with LVs for W well within the range of entangled states. Currently, we know that a LV model is possible for $W \lesssim 0.6829$ (Hirsch *et al.*, 2017) while nonlocality is guaranteed for $W \gtrsim 0.7056$ (Vértesi, 2008), only slightly better than the bound is immediately derivable from CHSH.[6]

As a proof, we present a simple rendering of Werner's original argument by Popescu (1994), which covers the range $W \leq \frac{1}{2}$. It is sufficient to find the LV model for $W = \frac{1}{2}$, since any ρ_W with $W < \frac{1}{2}$ can be obtained by mixing $\rho_{W=\frac{1}{2}}$ with white noise. In each round, Alice and Bob pre-share a classical variable in the form of a vector $\hat{\lambda}$ drawn from the surface of the unit sphere $\mathbb{S}^2$ with uniform distribution $\rho(\hat{\lambda})d\hat{\lambda} = \frac{1}{4\pi}\sin\theta d\theta d\varphi$ with the usual spherical coordinates. Alice draws her output by simulating the statistics of a measurement of a single spin prepared in the direction $\hat{\lambda}$:

---

[6] The value of the CHSH parameter for a Werner state is $2\sqrt{2}W$, which implies nonlocality for $W > 1/\sqrt{2} \approx 0.7071$. Vértesi proved that the reported minor improvement can be achieved, but one has to use inequalities with $M_A, M_B \gtrsim 465$ inputs per player!

$$P_{\hat{\lambda}}^A(a|\hat{a}) = \frac{1}{2}\left(1 + a\hat{a}\cdot\hat{\lambda}\right).$$ (3.25)

Bob outputs deterministically $b = -\text{sign}(\hat{b}\cdot\hat{\lambda})$, i.e., $b = +1$ if $\hat{b}\cdot\hat{\lambda} \leq 0$, $b = -1$ if $\hat{b}\cdot\hat{\lambda} > 0$). Thus

$$P(a, +1|\hat{a}, \hat{b}) = \int_{S^2} d\hat{\lambda}\rho(\hat{\lambda})\, P_{\hat{\lambda}}^A(a|\hat{a})\delta_{\hat{b}\cdot\hat{\lambda}\leq 0} = \frac{1}{4} + \frac{1}{2}a\int_{\hat{b}\cdot\hat{\lambda}\leq 0} d\hat{\lambda}\rho(\hat{\lambda})\hat{a}\cdot\hat{\lambda}.$$ (3.26)

In order to compute the integral, we choose the spherical coordinates such that $\hat{b}$ is $\hat{z}$ (i.e., $\theta = 0$):

$$\int_{\hat{b}\cdot\hat{\lambda}\leq 0} d\hat{\lambda}\rho(\hat{\lambda})\hat{a}\cdot\hat{\lambda} = \frac{1}{4\pi}\int_{\pi/2}^{\pi} d\theta\sin\theta\int_0^{2\pi} d\varphi\left[(a_{\hat{x}}\cos\varphi + a_{\hat{y}}\sin\varphi)\sin\theta + a_{\hat{z}}\cos\theta\right]$$

$$= \frac{1}{2}a_{\hat{z}}\underbrace{\int_{\pi/2}^{\pi} d\theta\sin\theta\cos\theta}_{=-\frac{1}{2}} = -\frac{1}{4}a_{\hat{z}}.$$

Inserting this result into (3.26) and recalling that $a_{\hat{z}} = \hat{a}\cdot\hat{b}$, we recover indeed (3.24) for $b = +1$. The calculation for $b = -1$ changes only in the bounds of the last integral and yields the desired result too.

The state-behavior (3.24) does not cover all possible statistics that can be obtained by measuring ρ_W, because it is restricted to projective measurements. The idea of adding POVMs to the state-behavior was first introduced by Barrett (2002), who found a LV description up to $W \leq \frac{5}{12} \approx 0.42$. At the moment of writing, this has been extended up to $W \lesssim \frac{2}{3} \times 0.6829 \approx 0.455$ (Hirsch et al., 2017; Oszmaniec et al., 2017). It is not yet known whether POVMs actually lead to nonlocal behaviors in the range $0.455 \lesssim W \lesssim 0.6829$.

3.3.2 Many-copy nonlocality

We start with the result that is simplest to state: There exist states ρ that have a LV model, but such that $\rho^{\otimes N}$ is a nonlocal resource without further processing. This has been called *super-activation of nonlocality*. It was first found in network configurations (i.e., with the N copies of the bipartite ρ shared among more than two players), but the most elegant result proves it for only two players (Palazuelos, 2012). It is slightly too technical to be presented here, and the reader can refer to the original reference that is self-contained.

Now, if Alice and Bob share $\rho^{\otimes N}$, they can submit this resource to some processing: Any processing performed *before receiving the inputs* is valid, as it constitutes merely the preparation of a new resource. Two types of interesting processes[7] have been considered.

The first is *entanglement distillation*: If the state ρ is distillable, then, given sufficiently many copies, with local operations and classical communication Alice and Bob can come arbitrarily close to sharing a maximally entangled state. Thus, any distillable state is a nonlocal resource for many-copy processing (Peres, 1996): In particular, so are all entangled Werner states. As for non-distillable states, their nonlocality has been elusive, with many negative results leading to the conjecture that no such state could be a nonlocal resource ("Peres' conjecture"). Eventually, Vértesi and Brunner (2014) refuted this conjecture by providing an example of a bound entangled state that is nonlocal in a single-copy setting.

The second is *filtering*, also known as "*revealing hidden nonlocality*" from the title of the paper that introduced it (Popescu, 1995). Considered the Werner states for higher dimension

$$\rho_{W'}^{(d)} = W' \frac{2\Pi_{\text{anti}}}{d(d-1)} + (1 - W') \frac{\mathbb{I}_d}{d^2} \tag{3.27}$$

where Π_{anti} is the projector on the antisymmetric subspace. Werner (1989) had proved that these states are local for projective measurements if $W' \leq \frac{d-1}{d}$. The filtering we consider is the projection, both on Alice and on Bob, in a two-dimensional subspace: $F_A = F_B = |0\rangle\langle 0| + |1\rangle\langle 1|$. The unnormalized state resulting from a successful filtering reads $F_A F_B \rho_{W'}^{(d)} F_A F_B = W' \frac{2|\Psi^-\rangle\langle\Psi^-|}{d(d-1)} + (1 - W') \frac{\mathbb{I}_2}{d^2}$: The normalized state is therefore a two-qubit Werner state (3.23) with

$$W = \frac{1}{1 + 2\frac{d-1}{d}\frac{1-W'}{W'}}. \tag{3.28}$$

Let us then set $W' = \frac{d-1}{d}$: The initial two-qudit state is local, but the filtered state violates CHSH if $d > \frac{2}{\sqrt{2}-1} \approx 4.8$ (i.e., for all $d \geq 5$). Notice that the filter is applied to each copy separately, but when it fails that copy is discarded: Thus, as a preparation of a resource, it is indeed a many-copy processing.

Popescu's proof considers initial states whose locality is proved only for projective measurements. Later, the activation of nonlocality through filtering was also proved for initial states, whose state-behavior is local for the most general local measurements, including sequential ones (Hirsch *et al.*, 2013). Also, an extension of Popescu's hidden nonlocality idea provides the only example known to date, in which *any* bipartite entangled state leads to nonlocality: If ρ is a bipartite entangled state, there exists an

---

[7] A process allowed by quantum theory is: Throw the state away and replace it with a better one. Obviously, it would work ...

entangled bipartite state $\rho\prime$ that does not violate CHSH even after filtering, such that $\rho \otimes \rho\prime$ violates CHSH after filtering (Masanes, Liang, and Doherty, 2008).

....

EXERCISES

Exercise 3.1 *Prove that the state-behavior associated to all projective measurements on a pure, non-maximally entangled two qubit state (3.15) is*

$$P(a, b|\hat{a}, \hat{b}) = \frac{1}{4}\left\{1 + c(aa_{\hat{z}} + bb_{\hat{z}}) + ab[a_{\hat{z}}b_{\hat{z}} + s(a_{\hat{x}}b_{\hat{x}} - a_{\hat{y}}b_{\hat{y}})]\right\} \tag{3.29}$$

where $c = \cos 2\theta$, $s = \sin 2\theta$, and $\hat{a}, \hat{b}$ are the directions associated to the measurements.

Exercise 3.2 *Derive the statistics (3.9) for the results of projective measurements on the singlet state. Hint: Though not necessary, you may exploit the fact that the singlet state is invariant under bilateral rotation.*

4

Review of Bipartite Bell Scenarios

Tout ne va pas aussi mal que si ça allait pire.
Things are not as bad as if they were worse.
Attributed to Coluche

This chapter reviews the main examples of bipartite Bell scenarios beyond the $(2, 2; 2, 2)$ scenario.

4.1 Unanticipated Complexity

Our knowledge of the $(2, 2; 2, 2)$ Bell scenario paints a simple picture: Up to obvious relabelings, there is only one Bell inequality, CHSH; its maximal violation in quantum theory is rather easily computed, and is achieved by suitable measurements on the maximally entangled state of two qubits. One may hope that other Bell scenarios can also be studied thoroughly with relative ease and confirm the simplicity of the picture. Unfortunately, that is not the case. Here is a sample of what may, and actually does, happen:

- When moving away from $(2, 2; 2, 2)$, the number of inequivalent Bell inequalities grows quickly, and their quantum bounds are usually not as straightforward to obtain as the Tsirelson bound.

- A frequent abuse of language, inspired by the CHSH intuition and perpetuated by the unfortunate title of some pioneering papers, consists in calling "Bell inequalities for qudits" the inequalities of a Bell scenario with $m = d$ outputs per player. The implicit idea is that projective measurements on d-dimensional systems should exhaust what quantum theory can achieve with d outputs. However, this proves to be wrong in general: The maximal violation of several inequalities requires systems with dimension strictly larger than the alphabet of the outputs.

- For several inequalities, the maximal quantum violation is achieved by measurements on a non-maximally entangled state. This is arguably the most counterintuitive discovery in the quantum theory of Bell nonlocality.[1]

---

[1] It is known from entanglement theory that one can deterministically create any state starting from the maximally entangled one, using local operations and classical communication. Thus, if the players can produce the maximally entangled state, they can always process it into the state of their choice prior to receiving the reviewer's queries.

Bell Nonlocality. Valerio Scarani. © Valerio Scarani 2019. Published in 2019 by Oxford University Press.
DOI: 10.1093/oso/9780198788416.001.0001

- Bell scenarios with larger alphabets contain all those with smaller alphabets; but, it is not *a priori* known if, in order to answer a given question, considering larger scenarios is useful or even needed. For instance, the nonlocality of arbitrary pure states can be assessed with CHSH and its simple variations (section 3.2.4), and no significant improvement on the CHSH threshold for the detection loophole was found (Appendix B.3). New inequalities may be "relevant" if they are violated by some mixed states, whose nonlocality was missed by all previously known inequalities. Only a few such examples are known[2] and there seems to be no systematic way of addressing this question.

In what follows, a large number of results are merely stated. In some cases, the proofs are very specific to each case and the interested reader can directly refer to the relevant article. Other results can be derived with tools that I have chosen to present in Part II of this book, namely the Navascués-Pironio-Acín (NPA) hierarchy of approximations (chapter 6) and the techniques of self-testing (chapter 7).

4.2 Several Inputs, Two Outputs

Among the Bell scenarios $(M, 2; M, 2)$ with $M > 2$, the local polytope has been fully solved only for $M = 3$ (see next) and $M = 4$ (Zambrini, Cruzeiro, and Gisin, 2019). For $M = 5$ up to 10, large incomplete lists of inequalities can be found in the literature, but to the best of my knowledge none is worth stressing in this book.

4.2.1 The (3, 2; 3, 2) scenario and the inequality I_{3322}

The local polytope for the $(3, 2; 3, 2)$ Bell scenario was first solved by Froissart (1981), then again several years later (Pitowsky and Svozil, 2001; Śliwa, 2003; Collins and Gisin, 2004). It has 684 facets. We expect to find 36 positivity facets. We also expect a certain number of liftings of CHSH: Specifically, Alice and Bob have each 3 ways to choose two inputs out of three, and the relabeling symmetries will generate 8 versions for each such choice, so we expect $3 \times 3 \times 8 = 72$ CHSH-like facets. Indeed, all these are found. The remaining 576 non-trivial facets are all equivalent, up to relabeling, to a single new inequality, denoted I_{3322}. In the Collins-Gisin representation, one of its versions reads

$$\mathcal{I}_{3322} \cdot \mathcal{P} \leq 0 \quad \text{with} \quad \mathcal{I}_{3322} = \begin{array}{c|c|c|c|} & -1 & 0 & 0 \\ \hline -2 & 1 & 1 & 1 \\ \hline -1 & 1 & 1 & -1 \\ \hline 0 & 1 & -1 & 0 \\ \end{array}. \tag{4.1}$$

---

[2] The first was the inequality I_{3322} that is presented in subsection 4.2.1. In subsection 3.3.1 we have already mentioned Vértesi's striking discovery that one has to go to $M_A, M_B \gtrsim 465$ in order to detect the nonlocality of some Werner states (Vértesi, 2008).

There is a lot to say about this simple inequality. First, contrary to CHSH, it cannot be rewritten in a way that involves only the correlation coefficients E_{xy}—in compact jargon, it is not a "correlation inequality." It is mildly surprising that an optimal criterion for a Bell test takes into account marginal information in a non-trivial way, given that what is hard for the players to generate are the correlations, not their local biases.

Much more surprising is the behavior of the maximal quantum violation. With measurements on two-qubit states, at most $\mathcal{I}_{3322} \cdot \mathcal{P} = \frac{1}{4}$ can be achieved, for suitable measurements on the maximally entangled state (Moroder *et al.*, 2013). Extensive numerical searches did not find any improvement using states of two qudits up to $d = 11$—and suddenly, for $d = 12$, higher violations are found, approaching the upper bound $\mathcal{I}_{3322} \cdot \mathcal{P} \lesssim 0.25087538$ set by the NPA hierarchy (Pál and Vértesi, 2010). An analytical optimization is still elusive at the moment of writing, but it is conjectured that the maximum be reached only in the limit $d \to \infty$.

Further, Collins and Gisin (2004) reported an example of a family of mixed states that certainly don't violate CHSH (as this can be tested using the criterion described in subsection 3.2.3) but that do violate I_{3322}. It is the first example beyond CHSH of the relevance of a Bell inequality for the detection of the nonlocality of quantum states. As the reader probably expects by now, nobody has been able to parametrize all the states that violate I_{3322}, not even for two qubits. This limitation hinders the study of the relevance of inequalities with even larger number of inputs.

4.2.2 The Mermin outreach criterion

In subsection 1.4.1, we presented Mermin's outreach criterion, a Bell test in the $(3, 2; 3, 2)$ scenario relying strongly on some correlations being perfect. The behavior

$$
\mathcal{P} = \begin{array}{|c||c|c|c|}
\hline
 & 1/2 & 1/2 & 1/2 \\
\hline\hline
1/2 & 1/2 & 1/8 & 1/8 \\
\hline
1/2 & 1/8 & 1/2 & 1/8 \\
\hline
1/2 & 1/8 & 1/8 & 1/2 \\
\hline
\end{array}, \tag{4.2}
$$

for which indeed $P(a_i = b_i) = 1$ and $P(a_i = b_{j \neq i}) = 1/4$, gives $O = 4.5$, which can be proved to be the quantum minimum (Exercise 6.2). This behavior may be obtained by measuring the state $|\Phi^+\rangle$ along the directions $\hat{a}_1 = \hat{b}_1 = \hat{z}$, $\hat{a}_2 = \hat{b}_2 = \frac{\sqrt{3}}{2}\hat{x} - \frac{1}{2}\hat{z}$ and $\hat{a}_3 = \hat{b}_3 = -\frac{\sqrt{3}}{2}\hat{x} - \frac{1}{2}\hat{z}$.

It is straightforward to check that this behavior does not violate the version (4.1) of I_{3322}, but this may be simply due to a mismatch in the choice of the labeling: one should check if any of the 576 versions is violated. As it turns out, the behavior (4.2) does not violate any version of I_{3322}, while it violates some versions of CHSH without achieving the Tsirelson bound (Exercise 4.1). Thus, in a study based on the local polytope, nothing similar to the Mermin outreach criterion appears as a tight inequality, nor does the behavior that maximises its violation play any special role.

4.2.3   Chained inequalities

Another family of inequalities that was proposed on intuitive grounds are the so-called *chained inequalities*, one for each $(M, 2; M, 2)$ Bell scenario (Pearle, 1970; Braunstein and Caves, 1990). The most transparent form of the chained inequality for M inputs is

$$C_M = P(a_1 = b_1) + P(b_1 = a_2) + P(a_2 = b_2) + \ldots$$
$$\ldots + P(a_M = b_M) + P(b_M \neq a_1) \leq 2M - 1. \tag{4.3}$$

This is a sum of $2M$ probabilities, which can't all be 1 for local behaviors, because if $a_1 = b_1 = a_2 = \ldots = b_M$, then $b_M = a_1$. Since the maximum can be reached with LD behaviors, for which probabilities are either 0 or 1, the bound of $2M - 1$ is indeed the highest we can expect. All the LD behaviors that saturate the bound are easy to list: There are the two behaviors $a_1 = \ldots = b_M = \pm 1$, that set $P(b_M \neq a_1) = 0$; the $2M$ behaviors $a_1 = \ldots = a_k \neq b_k = \ldots = b_M = \pm 1$, that set $P(a_k = b_k) = 0$, for $k \in \{1, \ldots, M\}$; and the $2(M-1)$ behaviors $a_1 = \ldots = b_k \neq a_{k+1} = \ldots = b_M = \pm 1$, that set $P(b_k = a_{k+1}) = 0$, for $k \in \{1, \ldots, M-1\}$. This counting also shows that (4.3) does not define a facet of the local polytope for $M > 2$: The inequality is saturated by only $4M$ LD behaviors, whereas there must be at least $D_{NS} = M^2 + 2M$ extremal points on each facet.

The most interesting feature of this family of inequalities is the fact that the quantum violation comes arbitrarily close to the algebraic limit $2M$ in the limit $M \to \infty$. Indeed, consider a two-qubit maximally entangled state

$$|\Phi^+\rangle = \frac{1}{\sqrt{2}}(|+\hat{z}\rangle|+\hat{z}\rangle + |-\hat{z}\rangle|-\hat{z}\rangle) \tag{4.4}$$

and measurements defined in the $\hat{x} - \hat{z}$ plane of the Bloch sphere as

$$\hat{a}_j = \cos\theta_{2j-1}\hat{z} + \sin\theta_{2j-1}\hat{x}, \quad \hat{b}_j = \cos\theta_{2j}\hat{z} + \sin\theta_{2j}\hat{x} \tag{4.5}$$

where $\theta_k = \frac{k\pi}{2M}$. With this choice, all the probabilities in (4.3) become equal to $\frac{1}{2}(1 + \cos\frac{\pi}{2M}) = \cos^2\frac{\pi}{4M}$ and consequently

$$C_M = M\left(1 + \cos\frac{\pi}{2M}\right) \gtrsim 2M - \frac{\pi^2}{8M}. \tag{4.6}$$

This violation is the maximal achievable with quantum resources (Wehner, 2006).

Now, $2M$ is very close to $2M - 1$ in the limit of large M, so these inequalities become increasingly sensitive to noise (Exercise 4.2). Thus, the chained inequalities are not a tool of choice for the practical certification of nonlocality. However, from the convergence of the quantum bound to the maximum, theorists have been able to derive foundational observations on quantum theory (section 9.4) and its potential applications (section 11.5).

Before moving on, let us mention two other forms of the same inequality that are often found. First, one can invert all the terms of the equation using $P(a_j = b_k) + P(a_j \neq b_k) = 1$ and get

$$
\begin{aligned}
C'_M &= 2M - C_M \\
&= P(a_1 \neq b_1) + P(b_1 \neq a_2) + \ldots + P(a_M \neq b_M) + P(b_M = a_1) \geq 1,
\end{aligned}
\tag{4.7}
$$

the maximal violation being now the minimum algebraic value $C'_M = 0$ for all M. This expression has the pleasant feature that neither the local nor the algebraic bound depend on M. Second, since the chained inequalities are correlation inequalities, one can rewrite them using only correlators: Indeed, using $E_{jk} = 2P(a_j = b_k) - 1 = 1 - 2P(a_j \neq b_k)$, one has

$$
\begin{aligned}
C''_M &= 2(C_M - M) \\
&= E_{11} + E_{21} + E_{22} + E_{32} + \ldots + E_{MM} - E_{1M} \leq 2(M-1).
\end{aligned}
\tag{4.8}
$$

From this form, it is manifest that the chained inequality for $M = 2$ is CHSH.

4.3 Two Inputs, Several Outputs

We move now to Bell scenarios $(2, m; 2, m)$ and study the family of CGLMP inequalities (Collins *et al.*, 2002a). There is one such inequality per value of m, often denoted I_{22mm} or I_m, as we shall do in this section. They are all facets of the corresponding local polytope (Masanes, 2003b). For $m = 3$, I_3 is the only inequality added to the liftings of CHSH. For $m > 3$, the local polytope has not been fully solved, but it is known that there are other inequalities than those of the CGLMP family. Here we give first a detailed presentation of I_3, then an overview of the common features of all I_m.

4.3.1 CGLMP for $m = 3$

The CGLMP inequality for $m = 3$, which had been described independently in (Kaszlikowski *et al.*, 2002), can be constructed in an instructive way. Start from the sum $P(a_0 = b_0) + P(a_0 = b_1) + P(a_1 = b_0) + P(a_1 = b_1 \oplus 1)$, where $\oplus$ denotes summation modulo 3. This is the same kind of contradiction as CHSH: If $a_0 = b_0, a_0 = b_1$ and $a_1 = b_0$, then LVs enforce $a_1 = b_1$; but the fourth probability is for $a_1 = b_1 \oplus 1$, so the four terms of the sum cannot be all 1 under LV (it is easy to check on all the local deterministic behaviors that the maximum is actually 3). Clearly, any triple of conditions leads to a similar enforcement: For instance, $a_0 = b_0$, $a_0 = b_1$ and $a_1 = b_1 \oplus 1$ would enforce $a_1 = b_0 \oplus 1$ under LVs. Now, one can penalize LV even more by *subtracting* from the previous sum the probabilities of the conditions enforced by LVs on each of the four triples:

$$I_3 = P(a_0 = b_0) \qquad +P(a_0 = b_1) \qquad +P(a_1 = b_0) \qquad +P(a_1 = b_1 \oplus 1)$$
$$-P(a_0 = b_0 \oplus 2) \ -P(a_0 = b_1 \oplus 1) \ -P(a_1 = b_0 \oplus 1) \ -P(a_1 = b_1) \qquad \leq 2. \tag{4.9}$$

This is the version of I_3 presented in (Collins *et al.*, 2002a). Notice that, since we work in modulo 3, we could have replaced the condition $a_1 \neq b_1$ by $a_1 = b_1 \oplus 2$ instead of $a_1 = b_1 \oplus 1$. But this is merely one of the many relabelings we are by now familiar with,[3] leading to an equivalent version of the inequality.

Having obtained the inequality, both (Collins *et al.*, 2002a) and (Kaszlikowski *et al.*, 2002) made the natural assumption that the maximal violation would be obtained with the maximally entangled state of two qutrits, $|\Phi\rangle = \frac{1}{\sqrt{3}}(|00\rangle + |11\rangle + |22\rangle)$, and went on to find the value and the corresponding measurements. Shortly later, (Acín *et al.*, 2002) wrote down the Bell operator for those measurements and discovered that its highest eigenvalue is larger than the value obtained previously, the associated eigenvector being the non-maximally entangled state

$$|\Phi(\gamma)\rangle = \frac{1}{\sqrt{2+\gamma^2}}(|0\rangle|0\rangle + \gamma|1\rangle|1\rangle + |2\rangle|2\rangle) \text{ with } \gamma = (\sqrt{11} - \sqrt{3})/2. \tag{4.10}$$

To add to the surprise, changing the measurements did not bring any improvement: The optimal measurements are the same for both $|\Phi\rangle$ and $|\Phi(\gamma)\rangle$, but a larger violation is obtained for the latter. As the argument stood, it relied on using qutrits: It was later confirmed that the maximal violation of this inequality uniquely identifies $|\Phi(\gamma)\rangle$ in the sense of self-testing (Bancal *et al.*, 2015). Thus, the maximal violation of CGLMP for $m = 3$ is achieved only for a non-maximally entangled state.

4.3.2   CGLMP for any m

Generalizing (4.9), the form of I_m found in (Collins *et al.*, 2002a) contains only probabilities of the form $P(a = b \oplus \delta | x, y)$, $\oplus$ representing sum modulo m. Thus, each inequality of the CGLMP family is manifestly a correlation inequality. For $m > 3$, however, the coefficients of those probabilities become a bit cumbersome. This is why various authors have opted for different forms, that are more compact, even if the correlation character is no longer transparent. Frequent practitioners will probably need a translation manual to navigate the literature; for the scope of this book, I shall simply mention two forms.

The most compact form is the one of (Zohren and Gill, 2008): Assuming the labeling $a, b \in \{0, \ldots, m - 1\}$, it reads

$$I_m^{ZG} = P(a \leq b|0, 0) + P(a \geq b|0, 1) + P(a \geq b|1, 0) + P(a < b|1, 1) \leq 3 \tag{4.11}$$

---

[3] It is *not* the relabeling $b_1 \to b_1 \oplus 1$, because this would affect the condition $a_0 = b_1$ too. Rather, it is the relabeling that consists in exchanging the roles of Alice and Bob: Indeed, if $a_1 = b_1 \oplus 1$, then $b_1 = a_1 \oplus 2$.

where $P(a \le b|0,0) = \sum_{b=0}^{m-1} \sum_{a \le b} P(a,b|0,0)$ and similarly for the other terms. The Collins-Gisin form for the same version of the inequality is

$$I_m^{CG} = \mathcal{I}_m \cdot \mathcal{P} \le 0 \text{ with } \mathcal{I}_m = \begin{Vmatrix} & -1 & -1 & \cdots & -1 & 0 & 0 & \cdots & 0 \\ \hline -1 & 1 & 1 & \cdots & 1 & 1 & 0 & \cdots & 0 \\ -1 & 0 & 1 & \cdots & 1 & 1 & 1 & \cdots & 0 \\ \vdots & \vdots & \vdots & \ddots & \vdots & \vdots & & \ddots & \vdots \\ -1 & 0 & 0 & \cdots & 1 & 1 & 1 & \cdots & 1 \\ \hline 0 & 1 & 0 & \cdots & 0 & -1 & 0 & \cdots & 0 \\ 0 & 1 & 1 & \cdots & 0 & -1 & -1 & \cdots & 0 \\ \vdots & \vdots & \vdots & \ddots & \vdots & \vdots & & \ddots & \vdots \\ 0 & 1 & 1 & \cdots & 1 & -1 & -1 & \cdots & -1 \end{Vmatrix} \qquad (4.12)$$

where each of the four square blocs is of dimension $(m-1) \times (m-1)$. In this form, it is manifest that I_2 is CHSH expressed in the CH form (2.32). The two forms are simply related by $I_m^{ZG} = I_m^{CG} + 3$ (Exercise 4.3).

Turning to quantum theory, there is no known analytical expression for the values of the maximal violation of I_m as a function of m. For small values of m, Zohren and Gill (2008) obtained lower bounds by numerical optimization assuming states of two m-dimensional systems and projective measurements, that were later found to coincide, within numerical precision, with upper bounds obtained with the NPA hierarchy. The reported maximal violations increases with m, although they are bounded by 4 as is obvious from (4.11). The states that achieve the maximal violation are never maximally entangled for any $m > 2$. Specifically, denoting $\{|k\rangle \,|\, k = 0 \ldots m-1\}$ the Schmidt bases of the optimal state for both Alice and Bob, the optimal measurements are found to be projections in the bases

$$|a\rangle_x = \sum_{k=0}^{m-1} \frac{e^{i\frac{2\pi}{m} ak}}{\sqrt{m}} e^{ik\alpha_x} |k\rangle, \quad |b\rangle_y = \sum_{k=0}^{m-1} \frac{e^{i\frac{2\pi}{m} bk}}{\sqrt{m}} e^{ik\beta_y} |k\rangle \qquad (4.13)$$

for $a, b \in \{0, 1, \ldots, m-1\}$, with $\alpha_0 = 0$, $\alpha_1 = \frac{\pi}{m}$, $\beta_0 = -\frac{\pi}{2m}$ and $\beta_1 = \frac{\pi}{2m}$.

4.4 Hardy's Test and the Magic Square

We finish this chapter by a detailed study of the two Bell tests based on extreme correlations that we introduced in section 1.4.

4.4.1 Hardy's test

Hardy's test was described in subsection 1.4.3. It is a test in the (2, 2; 2, 2) Bell scenario, so we know that the only tight inequality is CHSH. Like Mermin's outreach criterion, it is a test valid under some constraints on the behavior.

We work with pure two-qubit states. The calculation is cumbersome if one starts from the Schmidt decomposition (3.15), see Appendix G.2. A much more convenient form is

$$|\psi\rangle = \alpha\,|+\hat{z}\rangle\,|-\hat{z}\rangle + \alpha\,|-\hat{z}\rangle\,|+\hat{z}\rangle + \beta\,|-\hat{z}\rangle\,|-\hat{z}\rangle \qquad (4.14)$$

with $\alpha, \beta \in \mathbb{R}^+$ and $2\alpha^2 + \beta^2 = 1$ (Exercise 4.4). In this notation, the first constraint (1.11) is immediately satisfied for $A_0 = B_0 = \sigma_{\hat{z}}$. The second constraint (1.12) imposes that $P(b_1 = +1 | a_0 = -1) = 0$. We know that $|a_0 = -1\rangle \equiv |-\hat{z}\rangle$ of Alice, conditioned on which Bob's non-normalized state is $\alpha\,|+\hat{z}\rangle + \beta\,|-\hat{z}\rangle$. Therefore, this must be the eigenstate of B_1 for eigenvalue $b_1 = -1$. A symmetric argument can be made for the third constraint (1.13). Thus we have found

$$|a_1 = +1\rangle = |b_1 = +1\rangle = \frac{\beta}{\sqrt{\alpha^2 + \beta^2}}\,|+\hat{z}\rangle - \frac{\alpha}{\sqrt{\alpha^2 + \beta^2}}\,|-\hat{z}\rangle \qquad (4.15)$$

whence follows

$$P(+1, +1 | 1, 1) = \alpha^4 (1 - 2\alpha^2)/(1 - \alpha^2)^2. \qquad (4.16)$$

The LV prediction (1.14) is that this probability should be 0. This is the case only for $\alpha = 0$ (product state) or $\beta = 0$ (maximally entangled state), while every pure non-maximally entangled state of two qubits shows nonlocality in Hardy's test. The easiest example to remember is the case of $\alpha = \beta = \frac{1}{\sqrt{3}}$, for which $A_1 = B_1 = -\sigma_x$ and $P(+, +|1, 1) = \frac{1}{12}$ (Exercise 4.4). The maximal achievable value is

$$\max_{\alpha} P(+, +|1, 1) = \frac{5\sqrt{5} - 11}{2} \approx 0.0902 \qquad (4.17)$$

obtained for $\alpha = (\sqrt{5} - 1)/2$, that is for $A_1 = B_1 = (2 - \sqrt{5})\sigma_{\hat{z}} - 2\sqrt{\sqrt{5} - 2}\sigma_{\hat{x}}$. This value is the quantum maximum irrespective of the dimension and can only be reached with this two-qubit state in the sense of self-testing (Rabelo *et al.*, 2012). Our presentation of the Hardy's test violates CHSH in the form $- E_{00} + E_{01} + E_{10} + E_{11} \leq 2$, up to a value ≈ 2.3607 for the behavior that achieves (4.17).

It is instructive to understand why the maximally entangled state is not detected by Hardy's test. Denote by $|\psi_{B|a_0}\rangle$ the state on Bob's side conditioned on Alice having observed a_0. For maximally entangled states, $|\psi_{B|a_0=+1}\rangle$ and $|\psi_{B|a_0=-1}\rangle$ are orthogonal states (in particular, they belong to the same basis). Thus, the first two constraints $P(a_0 = +1, b_0 = +1) = 0$ and $P(a_0 = -1, b_1 = +1) = 0$ imply that $B_1 = -B_0$, and we have seen in subsection 3.1.3 that nonlocality cannot be demonstrated when one party's measurements commute. Conversely, $|\psi_{B|a_0=+1}\rangle$ and $|\psi_{B|a_0=-1}\rangle$ are generically *not* orthogonal for non-maximally entangled state. The iteration of the construction is known as *Hardy's ladder*.

4.4.2   The Magic Square

The Magic Square test was described in subsection 1.4.4. It is defined in a $(3, 8; 3, 8)$ Bell scenario. To describe the quantum measurement that achieves the perfect statistics, let's consider the following square of operators:

$$
\begin{array}{|c|c|c|}
\hline
\mathbb{I} \otimes \sigma_{\hat{z}} & \sigma_{\hat{z}} \otimes \mathbb{I} & \sigma_{\hat{z}} \otimes \sigma_{\hat{z}} \\
\hline
\sigma_{\hat{x}} \otimes \mathbb{I} & \mathbb{I} \otimes \sigma_{\hat{x}} & \sigma_{\hat{x}} \otimes \sigma_{\hat{x}} \\
\hline
-\sigma_{\hat{x}} \otimes \sigma_{\hat{z}} & -\sigma_{\hat{z}} \otimes \sigma_{\hat{x}} & \sigma_{\hat{y}} \otimes \sigma_{\hat{y}} \\
\hline
\end{array}
. \tag{4.18}
$$

The three operators on each line and on each column commute. Besides, denoting by O_{ij} the operator in the i-th line and j-th column, it holds

$$
\prod_{j=1}^{3} O_{ij} = \mathbb{I} \otimes \mathbb{I} \text{ for all } i, \quad \prod_{i=1}^{3} O_{ij} = -\mathbb{I} \otimes \mathbb{I} \text{ for all } j. \tag{4.19}
$$

Thus, if Alice's input $x \in \{1, 2, 3\}$ tells her to measure the three operators listed in line x of the square, and Bob's input $y \in \{1, 2, 3\}$ tells him to measure the three operators listed in column y of the square, the first two of the three conditions (1.15) are automatically satisfied for any state. In order to satisfy the third condition, we need to find a four-qubit state such that Alice's and Bob's outputs are the same for the operator where the line and the column intersect, i.e., $\langle \Psi | O_{xy} \otimes O_{xy} | \Psi \rangle = 1$. The state that does the job is the product of two maximally entangled two-qubit states: For the operators as written, it is $|\Phi^+\rangle_{A_1 B_1} |\Phi^+\rangle_{A_2 B_2}$ with $|\Phi^+\rangle$ given in (4.4). Once again, it was later proved to be unique in the sense of self-testing (Wu *et al.*, 2016).

This Bell test has been formulated requiring perfect correlations. One could try and relax those requirements by formulating the inequality

$$
\sum_{xy} P \left(\prod_j a_x^j = +1, \prod_k b_y^k = -1, a_x^{j=y} = b_y^{k=x} \right) \leq 8. \tag{4.20}
$$

This inequality is not a facet of the local polytope in the $(3, 8; 3, 8)$ Bell scenario (Gisin *et al.*, 2007). However, the verifier could ask Alice to enforce $\prod_j a_x^j = +1$ and Bob to enforce $\prod_k b_y^k = -1$, by outputting their third bit according to the rule. Now Alice and Bob output only two bits, thus this modification defines a Bell test in the $(3, 4; 3, 4)$ Bell scenario. For it, the inequality

$$
\sum_{xy} P \left(a_x^{j=y} = b_y^{k=x} \middle| \prod_j a_x^j = +1, \prod_k b_y^k = -1 \right) \leq 8 \tag{4.21}
$$

does define a facet of the local polytope.

....

EXERCISES

Exercise 4.1 *We consider the behavior (4.2) :*

(a) *Prove that it violates some liftings of CHSH. Recall that the Collins-Gisin representation of CHSH is (2.32).*

(b) *Prove that it does not violate any of the 576 versions of I_{3322}. This can be done by brute force on a computer; we rather suggest to do it by hand by keeping the inequality fixed and applying the symmetries on this (very symmetric) behavior.*

Exercise 4.2 *Prove that, in order to violate the chained inequality (4.3) with a two-qubit Werner state (3.23), one needs $W \gtrsim 1 - \frac{1}{M}$.*

Exercise 4.3 *Prove that the two forms (4.11) and (4.12) of the CGLMP inequalities are related by $I_m^{ZG} = I_m^{CG} + 3$ for no-signaling behaviors. Does this equivalence also hold for signaling behaviors? (If yes, prove it; if not, you should be able to find a behavior that violates one but not the other.)*

Exercise 4.4 *This exercise proves two statements made in discussing Hardy's test (subsection 4.4.1):*

1. *Prove that any two-qubit pure state can be written as (4.14). Due to the uniqueness of the Schmidt decomposition (3.15), it is enough to prove that the largest eigenvalue of the reduced states can take any value between $\frac{1}{2}$ and 1.*

2. *Prove that the state $\frac{1}{\sqrt{3}}\left(\left|+\hat{z}\right\rangle\left|-\hat{z}\right\rangle + \left|-\hat{z}\right\rangle\left|+\hat{z}\right\rangle + \left|-\hat{z}\right\rangle\left|-\hat{z}\right\rangle\right)$ leads to $P(+,+|1,1) = \frac{1}{12}$ for $A_1 = B_1 = -\sigma_x$.*

5

Multipartite Bell Nonlocality

Non serve essere quindici in squadra se tutti in propria area.
No point of being fifteen in a team, if all stay in their penalty area.

Attributed to V. Boskov

We conclude Part I with an overview of Bell nonlocality in multipartite scenarios.

5.1 Definition and Systematic Results

5.1.1 Multipartite local behaviors

The definition of a multipartite local behavior is the natural extension of the bipartite case (2.16). Denoting by $x^{(i)}$ and $a^{(i)}$ the input and output of the i-th player, with $i = 1, \ldots, n$, a behavior is local if its probability distributions can be written

$$P(a^{(1)}, \ldots, a^{(n)} | x^{(1)}, \ldots, x^{(n)}) \overset{LV}{=} \int d\lambda Q(\lambda) P_\lambda(a^{(1)} | x^{(1)}) \ldots P_\lambda(a^{(n)} | x^{(n)}) \qquad (5.1)$$

for all inputs.

It is then straightforward to extend Fine's theorem: In particular, any local behavior can be seen as convex combination of local deterministic behaviors, the latter being the extremal points of the local polytope.

In this sense, there is nothing fundamentally new in the definition of locality for multipartite Bell tests (although nonlocality is richer, since there are various ways in which a behavior may fail to be local, see section 5.3). That being said, the *complexity* of the local polytope grows very quickly, to the point that few systematic studies are known: We review them in subsections 5.1.3 and 5.1.4. Before this, let us turn to quantum theory and prove a result that was left pending in subsection 3.2.4: All pure entangled states are nonlocal resources, even multipartite ones.

5.1.2 All pure entangled states are nonlocal resources

We want to prove that any pure state $|\Psi\rangle_{A^{(1)}, \ldots, A^{(n)}}$ with $n > 2$ is a nonlocal resource if it's entangled. We follow the proof of (Popescu and Rohrlich, 1992a). Without loss of

Bell Nonlocality. Valerio Scarani. © Valerio Scarani 2019. Published in 2019 by Oxford University Press.
DOI: 10.1093/oso/9780198788416.001.0001

generality, the state is taken to be entangled in all its subsystems: If the state were of the form $|\phi\rangle_{A^{(1)},\ldots,A^{(m)}} \otimes |\phi'\rangle_{A^{(m+1)},\ldots,A^{(n)}}$, we know already that correlations across the tensor product cannot exhibit nonlocality, so we would consider the two groups separately.

We first consider a Bell test for the first and the second players $A^{(1)} \equiv A$ and $A^{(2)} \equiv B$. Each of the other players is queried with $x^{(j)} = 0$ ($j = 3, \ldots, n$), performs the corresponding measurement $\mathcal{M}^{x^{(j)}=0}$ and sends the output to the verifier. Only when the verifier receives a particular string of outputs, say $(a^{(3)}, \ldots, a^{(n)}) = (+1, \ldots, +1)$, he queries Alice and Bob with a normal CHSH test with $x, y \in \{1, 2\}$. The multipartite state being entangled in all its parties, the measurement operators $\mathcal{M}^{x^{(j)}=0}$ can always be chosen such that the conditional pure state $|\psi\rangle_{AB} \propto \Pi_{+1}^{x^{(3)}=0} \ldots \Pi_{+1}^{x^{(n)}=0} |\Psi\rangle_{A^{(1)},\ldots,A^{(n)}}$ is entangled.[1] Then, Alice and Bob can choose the measurements associated to the inputs $x, y \in \{1, 2\}$ such that the resulting *bipartite* behavior

$$\left\{ P(a, b|x, y; x^{(j)} = 0, a^{(j)} = +1 \text{ for } j = 3, \ldots, n) \,|\, x, y \in \{1, 2\}, a, b \in \{-1, +1\} \right\} \quad (5.2)$$

is nonlocal. But then, the *multipartite* behavior

$$\mathcal{P}_{aux} = \{P(a, b, +1, \ldots, +1|x, y, 0, \ldots, 0) \,|\, x, y \in \{1, 2\}, a, b \in \{-1, +1\}\}$$

is nonlocal too: If it were of the form (5.1), the bipartite behavior (5.2) would be local.

This means that the Bell test just described for players 1 and 2 can be a subset of rounds of a bigger Bell test that assesses the multipartite behavior, and *any* multipartite behavior $\mathcal{P}$ will be nonlocal as soon as it contains $\mathcal{P}_{aux}$ as a sub-behavior. This concludes the proof that any pure entangled state is a nonlocal resource—a disappointingly simple and certainly non-optimal proof, insofar as it relies only on pairwise nonlocality, but sufficient for the purpose.

5.1.3   The simplest multipartite Bell scenario

The simplest multipartite Bell scenario is the (2, 2; 2, 2; 2, 2) scenario, that involves three players, each with binary input and output. We can anticipate that the local polytope will have positivity facets and liftings of CHSH in some multiplicity. But when the full characterization was reported (Pitowsky and Svozil, 2001; Śliwa, 2003), it showed an unexpected explosion in complexity. The local polytope has 53856 facets—for comparison, recall that the two-player scenario has 24. After sorting them out for relabeling, there remain 46 inequivalent types of facets: The positivity facets and the liftings of CHSH are indeed present, and so is the Mermin inequality (1.10) that we'll meet again in the next section. For a rather systematic study of all the others, the reader can refer to (López-Rosa *et al.*, 2016).

---

[1] The formal proof is cumbersome to write down: In fact, the formalization given in Popescu and Rohrlich (1992a) was incorrect and was fixed in Gachechiladze and Guehne (2017).

Here, we single out the only inequality that cannot be violated with quantum behaviors (Almeida *et al.*, 2010):

$$G = P(0, 0, 0|0, 0, 0) + P(1, 1, 0|0, 1, 1) + P(0, 1, 1|1, 0, 1) + P(1, 0, 1|1, 1, 0) \leq 1.$$
$$(5.3)$$

This inequality has been called *"guess-your-neighbor-input" (GYNI) inequality* because the four probabilities are of the form $P(a = y, b = z, c = x|x, y, z)$. The local bound is easy to check on LD behaviors: $P(0, 0, 0|0, 0, 0) = 1$ means $a_0 = b_0 = c_0 = 0$, and then $P(1, 1, 0|0, 1, 1) = 0$ because it implies $a_0 = 1$, and so on. It is also easy to prove that quantum theory cannot violate the inequality. Indeed, in quantum theory, there exist states and measurements such that $P(a, b, c|x, y, z) = \text{Tr}(\rho \Pi_a^x \Pi_b^y \Pi_c^z) \equiv \text{Tr}(\rho \Pi_{abc}^{xyz})$. Now, notice that the four projectors Π_{abc}^{xyz} that enter (5.3) are mutually orthogonal: Π_{000}^{000} is associated to output $a = 0$ for measurement $x = 0$, while Π_{110}^{011} is associated to output $a = 1$ for the same measurement $x = 0$, and so on. Thus, $\mathcal{G} = \Pi_{000}^{000} + \Pi_{110}^{011} + \Pi_{011}^{101} + \Pi_{101}^{110} \leq \mathbb{I}$ and therefore $G = \text{Tr}(\rho \mathcal{G}) \leq 1$ as claimed.

Outputting another player's input deterministically is an extreme form of signaling, hence surely one cannot reach the algebraic limit $G = 4$ with no-signaling behaviors. It was proved that no-signaling behaviors can violate up to $G = \frac{4}{3}$. The GYNI inequality is thus irrelevant for the purpose of certifying the nonlocality of quantum resources. However, it has inspired some approaches to single out quantum theory among no-signaling theories (section 10.4).

The complexity of the local polytope[2] in this simplest multipartite Bell scenario has probably discouraged similar studies for larger Bell scenarios.

5.1.4 All correlation inequalities for two inputs, two outputs, *n* players

The other known systematic result in multipartite nonlocality is the classification of all *correlation inequalities* for every (2, 2; 2, 2; ...; 2, 2) scenario with arbitrary number n of players. These inequalities are referred to as WWZB inequalities (Werner and Wolf, 2001a; Żukowski and Brukner, 2002). To present the result, let us use the notation $a_{x_j}^{(j)} \in \{-1, +1\}$ for all $j = 1, ..., n$. In this notation, the correlator $E_{x_1...x_n}$ is simply the average value of the product of the outputs:

$$E_{x_1...x_n} \equiv E_{\underline{x}} = \left\langle \prod_{j=1}^{n} a_{x_j}^{(j)} \right\rangle.$$
$$(5.4)$$

WWZB proved that every tight correlation inequality is of the form

---

[2] The later characterization of the corresponding no-signaling polytope, a notion that we shall introduce in chapter 9, made the picture even more appalling (Pironio, Bancal, and Scarani, 2011).

$$\sum_{\underline{s}\in\{-1,+1\}^n} S(\underline{s}) \sum_{\underline{x}\in\{0,1\}^n} \left(\prod_{j=1}^{n} s_j^{x_j}\right) E_{\underline{x}} \le 2^N \tag{5.5}$$

where $S(\underline{s})$ can take only the values $+1$ and -1. Thus, this defines 2^{2^n} inequalities. Clearly, there is some redundancy in the list. Once sorted out by symmetries, the following number of inequivalent inequalities is found: 2 for $n = 2$ (we knew that: Positivity and CHSH), 5 for $n = 3$ (positivity, CHSH, Mermin and two more; that is, the 41 others are not correlation inequalities), 39 for $n = 4$, then ... more than 17476 for $n = 5$. Upon presenting this, WW tersely conclude that "for $n \ge 5$ [...] listing all essentially different inequalities is not going to be useful." The conditions (5.5) can be summarized in one *non-linear* inequality for the correlators,[3]

$$\sum_{\underline{s}\in\{-1,+1\}^n} \left| \sum_{\underline{x}\in\{0,1\}^n} \left(\prod_{j=1}^{n} s_j^{x_j}\right) E_{\underline{x}} \right| \le 2^N \tag{5.6}$$

which has been useful in deriving some analytical results.

5.2   Examples of Multipartite Bell Inequalities

We are going to give some examples of multipartite Bell inequalities. We focus mostly on a sub-family of WWZB, featuring one inequality for every number n of players, usually called MABK from the names of Mermin (1990a), who first reported such a construction, and Ardehali (1992) and Belinsky and Klyshko (1993a) who improved it.[4]

5.2.1   Construction of the MABK family

The MABK inequalities $M_n \equiv \langle \mathcal{M}_n \rangle \le 1$ are defined by a recursive construction of the correlators. The starting point of the recurrence is $M_2 \le 1$ with

$$\mathcal{M}_2 \equiv \frac{1}{2}\left(a_0^{(1)} a_0^{(2)} + a_0^{(1)} a_1^{(2)} + a_1^{(1)} a_0^{(2)} - a_1^{(1)} a_1^{(2)} \right), \tag{5.7}$$

which is just CHSH renormalized for convenience. Then, given $\mathcal{M}_{n-1}$, we have

$$\mathcal{M}_n = \frac{1}{2}\left(\mathcal{M}_{n-1}(a_0^{(n)} + a_1^{(n)}) + \mathcal{M}_{n-1}'(a_0^{(n)} - a_1^{(n)}) \right) \le 1 \tag{5.8}$$

---

[3] I have written this nonlinear inequality in the notation of ZB. In WW, it is equation (12), and the commentary that follows it is arguably the cheekiest sentence in the field (my italics): "*Obviously*, this nonlinear inequality is nothing but the characterization of the hyperoctahedron in 2n dimensions as the unit sphere of the Banach space ℓ^1. From this *simple* characterization of [the set of local correlations] it might seem that our problem is essentially trivial. However, ...".

[4] The review of Belinsky and Klyshko (1993b) provides an interesting window on the early days of nonlocality. When it comes to this topic (section 5.1 of the paper), a paper by Roy and Singh (1991) is mentioned, which somehow did not make it among the names retained in the acronym.

where $\mathcal{M}'_{n-1}$ is the equivalent version of the inequality obtained from $\mathcal{M}_{n-1}$ by permuting the labels of all players' inputs. The local bound of (5.8) can be proved recursively. Assume it holds for $n-1$ players, that is, $\mathcal{M}_{n-1}$ and $\mathcal{M}'_{n-1}$ can only be $+1$ or -1 for LD behavior. Then, a LD behavior on the n-th player has the familiar feature: If $a_0^{(n)} = a_1^{(n)} = \pm 1$, we have $(a_0^{(n)} + a_1^{(n)}) = \pm 2$ and $(a_0^{(n)} - a_1^{(n)}) = 0$; if $a_0^{(n)} = -a_1^{(n)} = \pm 1$, we have $(a_0^{(n)} + a_1^{(n)}) = 0$ and $(a_0^{(n)} - a_1^{(n)}) = \pm 2$.

Explicitly, the first iteration gives

$$
\begin{aligned}
\mathcal{M}_3 &= \frac{1}{2}\left(\mathcal{M}_2(a_0^{(3)} + a_1^{(3)}) + \mathcal{M}'_2(a_0^{(3)} - a_1^{(3)})\right) \\
&= \frac{1}{2}\left(a_0^{(1)} a_0^{(2)} a_1^{(3)} + a_0^{(1)} a_1^{(2)} a_0^{(3)} + a_1^{(1)} a_0^{(2)} a_0^{(3)} - a_1^{(1)} a_1^{(2)} a_1^{(3)}\right)
\end{aligned}
$$

that is,

$$
\mathcal{M}_3 = \frac{1}{2}(E_{001} + E_{010} + E_{100} - E_{111}) \leq 1 \tag{5.9}
$$

and is the Mermin inequality (1.10) that we have encountered in the introduction, rescaled by a factor $\frac{1}{2}$. Notice how only 4 out of 8 correlators appear in the expression. The next iteration gives

$$
\begin{aligned}
\mathcal{M}_4 = \frac{1}{4}\big[&-E_{0000} + (E_{0001} + E_{0010} + E_{0100} + E_{1000}) \\
&+ (E_{0011} + E_{0101} + E_{1001} + E_{0110} + E_{1010} + E_{1100}) \\
&- (E_{0111} + E_{1011} + E_{1101} + E_{1110}) - E_{1111}\big] \leq 1.
\end{aligned} \tag{5.10}
$$

This time, all 16 correlators appear in the expression. The inequality for 5 players $\mathcal{M}_5$ has again 16 correlators instead of the possible 32 (Exercise 5.1). The general form is $\mathcal{M}_n = 2^{-\lfloor n \rfloor/2}$ [sum of $2^{\lfloor n \rfloor}$ correlators], each correlator taken with the suitable sign; so the algebraic maximum is $2^{\lfloor n \rfloor/2}$. Also, the versions thus constructed are invariant under any permutation of the n players (although obviously there exist versions that are equivalent under relabelings and do not have this property).

5.2.2 Maximal violation in quantum theory

In quantum theory, the $\mathcal{M}_n$ should be seen as Bell operators

$$
\mathbf{M}_n = \frac{1}{2}\left(\mathbf{M}_{n-1} \otimes (A_0^{(n)} + A_1^{(n)}) + \mathbf{M}'_{n-1} \otimes (A_0^{(n)} - A_1^{(n)})\right) \tag{5.11}
$$

where every $A_{x_j}^{(j)}$ is an operator with eigenvalues $+1$ and -1.

Let us first derive an upper bound on the quantum violation, by a recursive proof. Suppose we have the upper bound for $n-1$ players: $\langle \mathbf{M}_{n-1} \rangle, \langle \mathbf{M}'_{n-1} \rangle \leq Q_{n-1}$.

Set $\mathbf{M}_{n-1} = Q_{n-1}\mathbf{A}$ and $\mathbf{M}'_{n-1} = Q_{n-1}\mathbf{A}'$, so that $\mathbf{M}_n = \frac{Q_{n-1}}{2}\mathbf{C}$ with $\mathbf{C} = \mathbf{A} \otimes (A_0^{(n)} + A_1^{(n)}) + \mathbf{A}' \otimes (A_0^{(n)} - A_1^{(n)})$. Now, $-\mathbb{I} \leq \mathbf{A}, \mathbf{A}' \leq \mathbb{I}$. Thus, $\mathbf{C}$ is a CHSH operator: It can be written as (3.6) and the Tsirelson bound $||\mathbf{C}||_\infty \leq 2\sqrt{2}$ follows. Putting everything together, we have got $Q_n \leq \sqrt{2}Q_{n-1}$; since $Q_2 \leq \sqrt{2}$, we conclude

$$M_n \leq 2^{(n-1)/2} \text{ for quantum behaviors.} \tag{5.12}$$

For odd n, this bound coincides with the algebraic bound $2^{\lfloor n/2 \rfloor}$, while for even n it is smaller than it by a factor $\frac{1}{\sqrt{2}}$.

The bound (5.12) is attained, for any n, by suitable projective measurements on the n-qubit GHZ state

$$|GHZ_n\rangle = \frac{1}{\sqrt{2}}\left(|+\hat{z}\rangle^{\otimes n} + |-\hat{z}\rangle^{\otimes n}\right) \tag{5.13}$$

(and only by those, in the sense of self-testing, chapter 7). The proof for arbitrary n is too lengthy for our purposes (Belinsky and Klyshko, 1993*a*; Scarani and Gisin, 2001). Let us work out explicitly the case $n = 3$, thus completing the claims left pending in subsection 1.4.2. It holds

$$\sigma_{\hat{x}} \otimes \sigma_{\hat{x}} \otimes \sigma_{\hat{x}}|GHZ_3\rangle = |GHZ_3\rangle \text{ and } \sigma_{\hat{x}} \otimes \sigma_{\hat{y}} \otimes \sigma_{\hat{y}}|GHZ_3\rangle = -|GHZ_3\rangle \tag{5.14}$$

plus permutations of the latter; in fact, these four eigenvalue equations define $|GHZ_3\rangle$ uniquely.[5] Thus, by choosing $\hat{a}_0 = \hat{b}_0 = \hat{c}_0 = \hat{y}$ and $\hat{a}_1 = \hat{b}_1 = \hat{c}_1 = -\hat{x}$ we obtain $E_{001} = E_{010} = E_{100} = -E_{111} = 1$, that is $M_3 = 2$.

5.3 Various Scenarios of Multipartite Nonlocality

5.3.1 Richness of nonlocal scenarios

Let us begin with a parable (Figure 5.1, left). Alice, Bob, and Charlie are going to take part in a three-partite Bell test. Upon inspecting the venue, they realize that Bob's and Charlie's rooms, though reached through different hallways, are actually adjacent and separated by a very thin wall: They will be able to communicate during the game! Alice, however, will be properly isolated. Can they now cheat the verifier in believing that they have a three-partite nonlocal resource?

Let's suppose the verifier checks the Mermin inequality. By communicating, Bob and Charlie can produce the correlations $E_{00} = E_{01} = E_{10} = -E_{11} = \pm1$; they can have agreed in advance in which rounds to use which option, and instructed Alice

---

[5] The intuition of Mermin (1990*a*) in building the inequality (1.10) was precisely the set of eigenvalue equations 5.14, related to the notion of *stabilizer group* that we won't introduce here. This intuition carries over to the MABK inequalities for odd n, but not for even n. The same intuition was later used to construct inequalities tailored to other graph states (Scarani *et al.*, 2005; Gühne *et al.*, 2005).

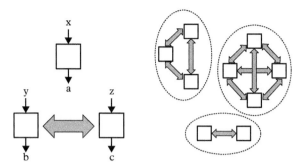

Figure 5.1 *The Svetlichny scenarios of multipartite nonlocality: Various groups correlated with LVs, with free communication inside each group. Left: The original scenario. Right: A possible multipartite scenario.*

to output $a_0 = a_1 = \pm 1$ accordingly. The verifier will then observe $M_3 = 2$, as well as all the marginals expected for the suitable measurements on the $|GHZ_3\rangle$ state. Thus, the Mermin inequality can be maximally violated if only two players share some nonlocal resource (in this case, communication), the third being classically correlated. One says that the Mermin inequality does not test *genuine* multipartite nonlocality. This observation was raised very early on by George Svetlichny (1987).

The parable illustrates a remark that we made at the start of the chapter: While the definition of locality is the obvious extension of the bipartite one, there are many ways in which a behavior may be nonlocal. The next subsection is devoted to Svetlichny scenarios, the following subsection to a generalization. For a synthetic classification of multipartite nonlocality scenarios, the reader can refer to (Chaves *et al.*, 2017).

5.3.2 Svetlichny scenarios

Svetlichny scenarios of multipartite nonlocality consist in grouping the players in several groups, such that communication is free within each group. Then one derives inequalities under the assumption that the different groups are correlated only through LVs (Figure 5.1). Svetlichny (1987) constructed the inequality for three parties,[6] the generalization to n parties followed several years later (Collins *et al.*, 2002b; Seevinck and Svetlichny, 2002).

In a compact presentation inspired by that of (Bancal *et al.*, 2011a), the Svetlichny-type expression for n players is built recursively as

$$\mathcal{S}_n = \mathcal{S}_{n-1} a_0^{(n)} + \mathcal{S}_{n-1}'' a_1^{(n)} \tag{5.15}$$

from $\mathcal{S}_1 = a_0^{(1)} + a_1^{(1)}$. The notation $\mathcal{S}_{n-1}''$ indicates the relabeling $a_0^{(1)} \to -a_1^{(1)}$ and $a_1^{(1)} \to a_0^{(1)}$ on the measurements and outcomes of only one player (here the first, but

---

[6] For the record, the paper has a typo in the expression of the inequality.

by symmetry it could be any). The first iteration gives $\mathcal{S}_2 = a_0^{(1)}a_0^{(2)} + a_1^{(1)}a_0^{(2)} - a_1^{(1)}a_1^{(2)} + a_0^{(1)}a_1^{(2)}$, that is CHSH; and $\mathcal{S}_2'' = -a_1^{(1)}a_0^{(2)} + a_0^{(1)}a_0^{(2)} - a_0^{(1)}a_1^{(2)} - a_1^{(1)}a_1^{(2)}$ is another version of CHSH.

The first interesting case is the three-partite case studied by Svetlichny. One way to read the expression $\mathcal{S}_3 = \mathcal{S}_2 a_0^{(3)} + \mathcal{S}_2'' a_1^{(3)}$ is the following: When the verifier queries Charlie with $z = 0$ (respectively, $z = 1$), he will test Alice and Bob with CHSH in the form $\mathcal{S}_2$ (respectively, $\mathcal{S}_2''$). If Alice and Bob share LVs, then both $\mathcal{S}_2$ and $\mathcal{S}_2''$ will be bounded by 2, and therefore

$$S_3 = \langle \mathcal{S}_3 \rangle \le 4, \tag{5.16}$$

with $S_3 = E_{000} + E_{100} + E_{010} - E_{110} - E_{101} + E_{001} - E_{011} - E_{111}$. This holds true even if Bob and Charlie can communicate freely, as in the previous parable. By symmetry, the reasoning is independent of which pair of players can communicate. Therefore, the Svetlichny inequality (5.16) holds for every behavior of the form

$$P(a, b, c|x, y, z) = \int d\lambda \, Q_c(\lambda) P_\lambda(a, b|x, y) P_\lambda(c|z)$$

$$+ \int d\mu \, Q_b(\mu) P_\mu(a, c|x, z) P_\mu(b|y)$$

$$+ \int d\nu \, Q_a(\nu) P_\nu(b, c|y, z) P_\nu(a|x), \tag{5.17}$$

where $\int d\lambda \, Q_c(\lambda) + \int d\mu \, Q_b(\mu) + \int d\nu \, Q_a(\nu) = 1$. These constraints define a polytope, called the Svetlichny polytope; the inequality (5.16) is one of its facets. In the (2, 2; 2, 2; 2, 2) scenario, none of the non-trivial facets of the local polytope is also a facet of this polytope (Pironio *et al.*, 2011).

To obtain the maximal quantum value, we can work in a similar way as we did for the MABK family: The Bell operator corresponding to $\mathcal{S}_3$ is

$$\mathbf{S}_3 = \mathbf{S}_2 \otimes A_0^{(3)} + \mathbf{S}''_2 \otimes A_1^{(3)} = \mathbf{A}_+ \otimes (A_0^{(3)} + A_1^{(3)}) + \mathbf{A}_- \otimes (A_0^{(3)} - A_1^{(3)})$$

with $\mathbf{A}_\pm = \frac{1}{2}(\mathbf{S}_2 \pm \mathbf{S}''_2)$. Since $-2\mathbb{I} \le \mathbf{A}_\pm \le 2\mathbb{I}$, we can achieve $\langle \mathbf{S}_3 \rangle = 4\sqrt{2}$ in violation of (5.16). This maximal violation is once again obtained for the $|GHZ_3\rangle$ state.

The case of $n > 3$ players follows the same pattern. For the Svetlichny bound, we can work by recursion: Suppose that

$$S_n = \langle \mathcal{S}_n \rangle \le 2^{n-1} \tag{5.18}$$

holds for any bipartition of the n players (we have just proved that it holds for $n = 3$). Then, when the verifier queries the $(n + 1)$-th player with $x^{(n+1)} = 0$ (respectively, $x^{(n+1)} = 1$), he will test the other n players with S_n (respectively, S_n''); whence $S_{n+1} \le 2^n$, confirming the recursion. The maximal quantum violation is attained by suitable measurements on the $|GHZ_n\rangle$ state (5.13) and is found to be $S_n = 2^{n-\frac{1}{2}}$, exceeding the two-groups bound (5.18) by a factor $\sqrt{2}$.

Other symmetries of the recursive definition (5.15), as well the relationship between the Svetlichny and the MABK family, are proposed as Exercise 5.2. In summary, some quantum behaviors exhibit genuine n-partite nonlocality in the sense of Svetlichny, for every n.

5.3.3 Scenarios with directional signaling

The Svetlichny definition of genuine multipartite nonlocality implies the possibility of *signaling* between some of the players. This has inspired two types of generalizations.

The first generalization consists in *using a refined classification of signaling behaviors*. In the definition (5.17) of the Svetlichny polytope, the two-player probabilities $P_\lambda(a, b|x, y)$, $P_\mu(a, c|x, z)$ and $P_\nu(b, c|y, z)$ are unrestricted. But the verifier is free to query the players at different times. For instance, getting back to our parable, the verifier may query Bob first, and query Charlie only after receiving Bob's answer. In this case, the only acceptable $P_\nu(b, c|y, z)$ are those in which Bob can communicate to Charlie but not the reverse. There may be behaviors that belong to the general Svetlichny polytope, but that cannot be decomposed on such directional behaviors: Their genuine multipartite nonlocality could then be detected with the suitable choice of timing[7] (Bancal *et al.*, 2013). Keeping directionality in view is also the best way to avoid formal mistakes when dealing with signaling behaviors (Gallego *et al.*, 2012): We sketch the pitfalls in which one can fall in Appendix G.3.

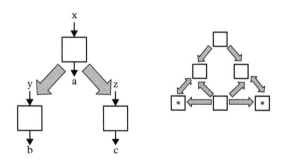

Figure 5.2 *Approaching multipartite nonlocality taking into account the directionality of communication. Jones, Linden, and Massar proved that Svetlichny's inequalities also hold as soon as there are two players in the network, such that no player knows both their inputs. Left: The simplest setting in which the players cannot be separated in non-communicating groups, and yet none of the three players gets to know both y and z. Right: A six-player scenario, in which nobody gets to know both the inputs of the two players denoted by a star.*

---

[7] For the record, what ultimately converged in the paper (Bancal *et al.*, 2013) was cited for several years as "Barrett and Pironio, in preparation".

The second generalization consists in considering *other scenarios with limited communication links* (Figure 5.2). Instead of allowing free communication inside non-communicating groups, communication can be restricted in its directionality: For instance, Alice can signal to both Bob and Charlie, while they cannot reply nor communicate between the themselves. Such a scenario is not captured by (5.17), but it turns out that Svetlichny's inequality still holds. In fact Jones, Linden, and Massar (2005) proved that in any *n*-player scenario, the bound (5.18) holds as soon as there is *at least one pair of players (i, j), such that no player knows both x_i and x_j* (for instance, in the example just put, no player knows both $y = x_B$ and $z = x_C$).

....

EXERCISES

Exercise 5.1 *Derive the MABK inequality for five players M_5, using the iterative definition given in subsection 5.2.1.*

Exercise 5.2 *In this exercise, we explore two consequences of the definition (5.15) of Svetlichny inequalities.*

1. *Prove by recursion that*

$$\mathcal{S}_n = \frac{1}{2}\left(\mathcal{S}_{n-k}\mathcal{S}_k + \mathcal{S}''_{n-k}\mathcal{S}''_k\right) \tag{5.19}$$

   is valid for all $k = 1, \ldots, n - 1$. Hint: Recall that the permutation '' affects only one player, and therefore $(\mathcal{S}_{n-k}\mathcal{S}_k)'' = \mathcal{S}''_{n-k}\mathcal{S}_k$.

2. *In (Collins et al., 2002b), the Svetlichny expression was defined in terms of the MABK expressions as follows:*

$$\mathcal{S}_n = \begin{cases} \mathcal{M}_n & \text{for } n \text{ even} \\ \frac{1}{2}(\mathcal{M}_n + \mathcal{M}'_n) & \text{for } n \text{ odd.} \end{cases} \tag{5.20}$$

   The definition (5.15) leads to the same, up to some relabelings and scaling. Show it for $n = 3$ and $n = 4$.

Part II

Nonlocality as a Tool for Certification

This second part presents Bell nonlocality as a tool for applied physics. The key observation is simple: On the one hand, nonlocality is assessed on the basis of the observed behavior alone; on the other hand, in the framework of quantum theory, nonlocality can only happen in the presence of entanglement, which itself is a resource for tasks like randomness generation and quantum key distribution. Therefore, the observation of nonlocality certifies entanglement (and possibly its usefulness) in a *device-independent* way, i.e., without characterization of the physical degrees of freedom.

6

The Set of Quantum Behaviors

Go, go, go, said the bird: human kind/
Cannot bear very much reality.

T.S. Eliot, *Burnt Norton*

After an introduction to device-independent certification that motivates the whole Part II, this chapter deals with the definition of the set of quantum behaviors and with some useful outer approximations to it.

6.1 Device-Independent Certification: A First Introduction

In a Bell test, the players should be on top of what they are doing—to say it like a physicist: The experimentalist must have a good control of the degrees of freedom and of the corresponding measurements. But none of this is needed by the verifier. Bell nonlocality can be certified *without any characterization of the devices*. We say that it is a *device-independent (DI) certification* of nonlocality.

Bell nonlocality has implications. We have seen that the only no-signaling deterministic strategies are the local deterministic ones (Exercise 2.2). Therefore, if the observed behavior is nonlocal and the resources are guaranteed to be no-signaling, the processes that produced the outputs are certainly *random*. One can then try and infer a bound on the amount of randomness from the certification of nonlocality. It would then be a DI certification of randomness: Something pretty disruptive, insofar as one is able to determine that a process is random without any modeling of the process itself. This is indeed possible, and we shall dedicate chapter 8 to it. Several other DI certifications can be based on nonlocality: Lower bounds on the secret key that can be created by quantum key distribution, on the amount of entanglement present in the system, on the dimensionality of the system... These are reviewed in Appendix F, alongside with the history of these discoveries and some confusions worth clarifying.

The theory of DI certification is challenging: One wants to assess (say) how much randomness is present by assuming that quantum theory is correct, but without being able to specify even the dimension of the Hilbert space of the system under study. Even brute force optimization becomes impossible. This calls for the characterization of the set of quantum behaviors that was postponed in chapter 3. This chapter is devoted to it.

Bell Nonlocality. Valerio Scarani. © Valerio Scarani 2019. Published in 2019 by Oxford University Press.
DOI: 10.1093/oso/9780198788416.001.0001

6.2   Definition and Geometry of the Quantum Set

6.2.1   The definition

Given a Bell scenario, the quantum set $\mathcal{Q}$ is defined as the set of all the quantum behaviors in that scenario, i.e., the behaviors that can be observed assuming that the players share quantum resources.

As for local behaviors, $\mathcal{Q}$ is first defined under the i.i.d. assumption; later we'll prove that the definition remains the same even if the underlying strategy is not i.i.d. Let us start by repeating the definition of a quantum behavior given in section 3.1 using the notations introduced there:

Definition 6.1 *A behavior $\mathcal{P}$ belongs to the quantum set $\mathcal{Q}$ of the $(M_A, m_A; M_B, m_B)$ Bell scenario if there exist:*

- *Two Hilbert spaces $\mathcal{H}_A$ and $\mathcal{H}_B$, of unspecified dimension, that could even be infinite*
- *A state ρ in the space of linear operators on $\mathcal{H}_A \otimes \mathcal{H}_B$*
- *A family of m_A-output measurements $\{\mathcal{M}_A^x | x \in \mathcal{X}\}$, whose elements Π_a^x are linear operators on $\mathcal{H}_A$*
- *A family of m_B-output measurements $\{\mathcal{M}_B^y | y \in \mathcal{Y}\}$, whose elements Π_b^y are linear operators on $\mathcal{H}_B$,*

such that

$$P(a, b|x, y) = Tr(\rho \Pi_a^x \otimes \Pi_b^y). \tag{6.1}$$

The extension to multipartite Bell scenarios is immediate.

The fact that the dimensions of the players' Hilbert spaces are left unspecified is important in several respects. First of all, it guarantees that the players' strategies are not restricted. In particular, the state ρ does not need to describe only the shared entanglement, it can include an arbitrary amount of local information as well.

Second, by Naimark's theorem, we can consider only *projective measurements* without loss of generality. By the same token, one could enlarge the Hilbert space to include the purification of the state and assume that the state is pure. We shall do it occasionally, notably for self-testing (chapter 7). However, as we shall see in chapter 8, for some adversarial tasks the purification may not be in the players' hands.

The third consequence is the convexity of the set, see subsection 6.2.3.

6.2.2   Another set, and the "Tsirelson problem"

In the development of tools to study the quantum set, a few rigoros mathematical results could be proved only by altering the definition of the set under study:

Definition 6.2 *A behavior $\mathcal{P}$ belongs to the set $\mathcal{Q}'$ of the $(M_A, m_A; M_B, m_B)$ Bell scenario if there exist:*

- *A Hilbert space $\mathcal{H}$, of unspecified dimension, possibly infinite*
- *A state ρ in the space of linear operators on $\mathcal{H}$*
- *A family of m_A-output measurements $\{\mathcal{M}_A^x | x \in \mathcal{X}\}$, and a family of m_B-output measurements $\{\mathcal{M}_B^y | y \in \mathcal{Y}\}$, whose elements are linear operators on $\mathcal{H}$ and satisfy*

$$[\Pi_a'^x, \Pi_b'^y] = 0 \text{ for all } x, a, y, b \tag{6.2}$$

such that

$$P(a, b|x, y) = \mathrm{Tr}(\rho \Pi_a'^x \Pi_b'^y). \tag{6.3}$$

In words, the requirement that Alice's and Bob's measurements are in tensor product is replaced by the requirement that all the operators in the family $\{\mathcal{M}_A^x | x \in \mathcal{X}\}$ commute with all the operators in the family $\{\mathcal{M}_B^y | y \in \mathcal{Y}\}$. It is clear that $\mathcal{Q} \subseteq \mathcal{Q}'$, because the representation $\Pi_a'^x = \Pi_a^x \otimes \mathbb{I}$ and $\Pi_b'^y = \mathbb{I} \otimes \Pi_b^y$ satisfies (6.2). The reverse implication would mean the following: When two complete sets of operators commute, it should be possible to ascribe different degrees of freedom to them, i.e., to find a tensor product structure. Tsirelson first claimed that this can be proved, but subsequently retracted the general claim, and this question became known as "Tsirelson's problem." For an introduction and early references, we refer to (Navascués *et al.*, 2012); the state of the art is best summarized in the introduction to (Coladangelo and Stark, 2018). In a nutshell: Whenever the behavior $\mathcal{P}$ can be obtained with finite-dimensional Hilbert spaces, the definition through the commutator implies the possibility of constructing a tensor product representation; and it is also known that $\mathcal{Q} = \mathcal{Q}'$ for the Bell scenario (2, 2; 2, 2) and a few other cases. However, in general $\mathcal{Q}$ is not a closed set (Slofstra, 2016): Whether its closure is either equal to, or strictly contained into, $\mathcal{Q}'$, is the remaining open question.[1]

In the following, we work mostly with (6.3) for simplicity of notation, but a tensor product can be read there too unless stated otherwise.

6.2.3 Geometry

The quantum set $\mathcal{Q}$ is a *convex* set. The proof relies on the fact that the Hilbert space dimensions are unspecified:

$$P(a,b|x,y) = \sum_k p_k \mathrm{Tr}(\rho_k \Pi_a^{x,k} \Pi_b^{y,k}) = \mathrm{Tr}(\rho \Pi_a^x \Pi_b^y)$$

---

[1] Though of mathematical importance, the fact that $\mathcal{Q}$ is not closed is irrelevant in practice, since every observed behavior will belong to the interior. However, if the closure $\mathcal{Q}$ were different from $\mathcal{Q}'$, this difference could be the object of tests: Indeed, both are convex sets, so the separating hyperplane theorem implies that there should a constant gap between them. A difference may have further profound implications: For instance, in the field of algebraic quantum field theory, localization is defined through commutators and not through tensor products. *After completion of this book, a proof was found that the two sets are not equivalent (Ji et al. 2020).*

with[2] $\rho = \bigoplus_k p_k \rho_k$, $\Pi_a^x = \bigoplus_k \Pi_a^{x,k}$ and $\Pi_b^y = \bigoplus_k \Pi_b^{y,k}$. Conversely, if one fixes the Hilbert space dimension of the shared resources, the set of achievable quantum behaviors is generally not convex,[3] although it may be in some cases (Donohue and Wolfe, 2015).

From convexity, it also follows that, if the underlying strategy is not i.i.d., the effective single-round behavior still belongs to $\mathcal{Q}$ as just defined. The proof follows the same pattern as the one for local behaviors given in subsection 2.6.1. Indeed, if

$$P(a_1, a_2, b_1, b_2 | x_1, x_2, y_1, y_2) = \mathrm{Tr}(\rho \Pi_{a_1 a_2}^{x_1 x_2} \Pi_{b_1 b_2}^{y_1 y_2}),$$

the corresponding effective single-round behavior reads

$$P'(a, b | x, y) = \frac{1}{2}\mathrm{Tr}(\rho E_a^x E_b^y) + \frac{1}{2}\mathrm{Tr}(\rho F_a^x F_b^y) \tag{6.4}$$

with the two POVMs

$$E_a^x E_b^y = \sum_{a_2, x_2} \Pi_{a a_2}^{x x_2} \sum_{b_2, y_2} \Pi_{b b_2}^{y y_2} \frac{P(x, x_2, y, y_2)}{P'(x, y)},$$

$$F_a^x F_b^y = \sum_{a_1, x_1} \Pi_{a_1 a}^{x_1 x} \sum_{b_1, y_1} \Pi_{b_1 b}^{y_1 y} \frac{P(x_1, x, y_1, y)}{P'(x, y)}.$$

Here, $P'(x, y) = \frac{1}{2}[\sum_{x_2, y_2} P(x, x_2, y, y_2) + \sum_{x_1, y_1} P(x_1 x, y_1, y)]$. In summary, $P'(a, b | x, y)$ is a convex sum of behaviors that belong to $\mathcal{Q}$, and thus belongs to $\mathcal{Q}$ as well.

Contrary to the set of local behaviors, $\mathcal{Q}$ is *not a polytope*: I continuously has many extremal points, even in the simplest Bell scenario. In other words, some of its boundaries are not hyperplanes, but convex curves, the most famous example of which is presented in the next subsection. The first attempts at systematic studies are uncovering a complex[4] zoology of geometric features (Goh *et al.*, 2018; Duarte *et al.*, 2018; Rai *et al.*, 2019). Overall, the only known way to prove that a behavior is on the boundary of $\mathcal{Q}$ consists in showing that it has a quantum realization while proving that it lies at the boundary of a superset (thus, on that point, the superset and the set coincide). Techniques to define supersets of $\mathcal{Q}$ are discussed in section 6.3, while finding a quantum realization basically requires a good guess.

6.2.4   An example of boundary

The most famous example of a curved boundary of $\mathcal{Q}$ is the boundary of the quantum set of the (2, 2; 2, 2) scenario in the slice with unbiased marginals $\langle a_0 \rangle = \langle a_1 \rangle = \langle b_0 \rangle =$

---

[2] One possible realization of these direct sums is simple: In each round, the players draw a value of k with probability p_k and act accordingly.

[3] Fixing the dimension may have consequences for the local set too. For instance, if the fixed local dimension of Alice's (Bob's) system is smaller than the number of outputs m_A (m_B), one cannot even produce all the local deterministic points of the Bell scenario.

[4] At least, as complex as allowed by the fact that the set is convex.

$\langle b_1 \rangle = 0$. Usually it is referred to as TLM boundary[5] (Tsirel'son, 1987; Landau, 1988; Masanes, 2003a). Above the CHSH facet (2.29), the equation of this boundary is

$$\text{Arcsin}(E_{00}) + \text{Arcsin}(E_{01}) + \text{Arcsin}(E_{10}) - \text{Arcsin}(E_{11}) = \pi; \qquad (6.5)$$

for the boundary above any of the other seven facets, one has to apply the corresponding permutations.

The TLM boundary defines a three-dimensional surface embedded in $\mathbb{R}^4$ (recall that $D_{NS} = 8$ for this Bell scenario). A possible parametrization is[6]

$$E_{xy} = \cos(\alpha_x - \beta_y) \text{ with } \alpha_1 \leq \beta_0 \leq \alpha_0 \leq \beta_1, |\alpha_x - \beta_y| \leq \pi. \qquad (6.6)$$

These correlations can be achieved by measuring the maximally entangled state of two qubits $|\Phi^+\rangle$ with measurements defined by the directions $\hat{a}_x = \cos\alpha_x \hat{z} + \sin\alpha_x \hat{x}$ and $\hat{b}_y = \cos\beta_y \hat{z} + \sin\beta_y \hat{x}$. This proves that the TLM curve belongs to $\mathcal{Q}$. The proof that it is also an upper bound will be sketched in subsection 6.3.3; alternatively, it can be inferred from the fact that the quantum realization just described is unique in the sense of self-testing, see chapter 7.

6.3 Semidefinite Relaxations of the Quantum Set

The quantum set is hard to characterize exactly, so it is often convenient to work with approximations. One obvious approximation consists in restricting the possible states and/or measurements, but the resulting set of behaviors is a subset of $\mathcal{Q}$. In view of device-independent certification, we'd rather have *supersets* (or *relaxations*) of $\mathcal{Q}$ so that, if something is proved for one of these supersets, it is automatically proved for the whole of the quantum set.

This section presents a construction of such supersets, whose membership is defined by compact and algorithmically efficient conditions. It was introduced in a systematic way by Navascués, Pironio, and Acín (2007; 2008) and thence referred to as *NPA relaxations*.

6.3.1 The construction

The NPA relaxations are based on the following observation:

Lemma 6.1 *Let $\mathcal{F} = \{F_1, \ldots, F_n\}$ be a collection of linear operators. Then, for any state ρ, the hermitian matrix $\Gamma(\rho, \mathcal{F})$ whose entries are*

---

[5] Lluís Masanes derived this bound during his Ph.D. thesis, unaware that it was already known. When told after posting his draft on the arXiv, he did not submit it for publication. Nonetheless, his derivation being different than the two others, the community added his initial to the acronym.

[6] Indeed, if $\theta \in [0, \pi]$, using $\cos\theta = \sin(\frac{\pi}{2} - \theta)$ we have $\text{Arcsin}(\cos\theta) = \frac{\pi}{2} - \theta$. With the order of angles chosen in (6.6) and exploiting $\cos\theta = \cos(-\theta)$, the l.h.s. of (6.5) reads $(\frac{\pi}{2} - \alpha_0 + \beta_0) + (\frac{\pi}{2} - \beta_1 + \alpha_0) + (\frac{\pi}{2} - \beta_0 + \alpha_1) - (\frac{\pi}{2} - \beta_1 + \alpha_1)$ which is indeed equal to π.

$$\Gamma_{ij} = \text{Tr}(\rho F_i^\dagger F_j) \tag{6.7}$$

is positive semidefinite.

Proof The proof is very direct: For any vector $\vec{v} \in \mathbb{C}^n$ it holds

$$\vec{v}^\dagger \Gamma \vec{v} = \text{Tr}\left[\rho \left(\sum_i v_i^* F_i^\dagger\right)\left(\sum_j v_j F_j\right)\right] \geq 0$$

because both ρ and any operator of the form $C^\dagger C$ are positive semidefinite.

Now, if $\mathcal{P} \in \mathcal{Q}$, we can consider the matrix $\Gamma_1 = \Gamma(\rho, \mathcal{F}_1)$ for the state ρ and the collection of $M_A m_A + M_B m_B$ projectors

$$\mathcal{F}_1 = \left\{\mathbb{I}, \{\Pi_a^x | x \in \mathcal{X}, a \in \mathcal{A}\}, \{\Pi_b^y | y \in \mathcal{Y}, b \in \mathcal{B}\}\right\}. \tag{6.8}$$

If we knew the state and the measurements, we could compute all the entries of Γ_1. In a device-independent approach, however, we don't know anything *a priori*, we are just assuming that the state and the projectors exist. Still, the value of *some* entries of Γ_1 is determined by the behavior, namely $\text{Tr}(\rho \Pi_a^x \Pi_b^y) \equiv P(a,b|x,y)$, $\text{Tr}(\rho \Pi_a^x) \equiv P(a|x)$, and $\text{Tr}(\rho \Pi_b^y) \equiv P(b|y)$. The other entries, involving two operators of Alice's or two of Bob's, do not correspond to anything observed in a Bell test, but Lemma 6.1 guarantees that they can be filled so as to have $\Gamma_1 \geq 0$. *A contrario*, suppose that after filling the suitable entries of Γ_1 with the behavior, one were to find that there is no way to fill the other entries to obtain $\Gamma_1 \geq 0$: Then surely $\mathcal{P} \notin \mathcal{Q}$.

Clearly, we have just described a *necessary* condition for a behavior $\mathcal{P}$ to belong to the quantum set $\mathcal{Q}$. For some parts of the quantum set, this condition is already sufficient: For instance, the TLM boundary (6.5) can be recovered, see subsection 6.3.3. In general, this condition is just the first of the *NPA hierarchy of criteria* (Navascués et al., 2007, Navascués et al., 2008). The set of behaviors that satisfy $\Gamma_1 \geq 0$ is denoted $\mathcal{Q}_1$.

For the next step of the hierarchy, one constructs the matrix Γ_2 for the collection

$$\mathcal{F}_2 = \mathcal{F}_1 \cup \left\{\left\{\Pi_a^x \Pi_{a'}^{x'}\right\}, \left\{\Pi_b^y \Pi_{b'}^{y'}\right\}, \left\{\Pi_a^x \Pi_b^y\right\}\right\} \tag{6.9}$$

where it is understood that all the new $(M_A m_A)^2 + (M_B m_B)^2 + M_A m_A M_B m_B$ operators should be added to $\mathcal{F}_1$. The set of behaviors for which $\Gamma_2 \geq 0$ is denoted $\mathcal{Q}_2$. The positivity of Γ_2 is a necessary condition for $\mathcal{P} \in \mathcal{Q}$, tighter than the positivity of Γ_1 but once again not sufficient in general. Notice that none of the entries involving operators like $\Pi_a^x \Pi_{a'}^{x'}$ will be determined by the behavior; nonetheless, such terms enforce some structure onto the matrix: For instance, $\text{Tr}(\rho(\Pi_a^x \Pi_{a'}^{x'})\Pi_{a''}^{x'}) = \text{Tr}(\rho \Pi_a^x \Pi_{a'}^{x'})\delta_{a',a''}$.

The reader has probably guessed rightly that the n-th step of the hierarchy involves constructing the matrix Γ_n for the collection of all products of n or fewer projectors. The set of behaviors for which $\Gamma_n \geq 0$ is denoted $\mathcal{Q}_n$. Clearly, $\mathcal{Q}_1 \supseteq \mathcal{Q}_2 \supseteq \mathcal{Q}_3 \supseteq \ldots$ and

$Q_n \supseteq Q'$ for all n. NPA proved that this hierarchy is convergent, i.e., $\lim_{n\to\infty} Q_n = Q'$. Notice that the convergence is proved for Q' and not for Q. Also, in many cases the quantum result is already obtained for a finite n.

The elegance of the hierarchy should not conceal the fact that Lemma 6.1 allows for *full freedom in choosing the collection of operators* $\mathcal{F}$. For instance, in section 10.5 we shall discuss an interesting role for the set of behavior Q_{1+AB}, obtained by only adding to $\mathcal{F}_1$ the operators of the form $\Pi_a^x \Pi_b^y$. Also it is good to keep in mind that adding a single operator to an existing $\mathcal{F}$ may be enough to obtain a tighter bound. We shall use the notation $Q_{\mathcal{F}}$ for the set of behaviors such that $\Gamma \geq 0$ for the collection of operators $\mathcal{F}$.

The fact that the NPA conditions are semidefinite requirements $\Gamma \geq 0$ is particularly appealing because it suggests the possibility of using them in *semidefinite programs* (SDPs). These are the next-to-simplest instance of convex optimization, after the linear programs that we encountered in chapter 2. Just as with linear programs, given a SDP, a dual SDP can be algorithmically constructed, and one can obtain simultaneously a lower and an upper bound to the desired solution. For more details on SDPs we refer to Appendix E and to the book of Boyd and Vandenberghe (2004). Now we are going to see three examples of such SDPs, more will be presented in the next chapters.

6.3.2   Example 1: Membership in the quantum set

Let's describe how the NPA techniques address the problem of membership, i.e., assessing whether $\mathcal{P} \in Q$. For any choice of $\mathcal{F}$, one can relax the membership problem to the following:

$$\Lambda = \max \lambda$$
$$\text{subject to} \quad \Gamma - \lambda \mathbb{I} \geq 0$$
$$\Gamma_{(x,a)} = P(a|x), \Gamma_{(y,b)} = P(b|y), \Gamma_{(x,a)(y,b)} = P(a, b|x, y) \tag{6.10}$$

with the obvious definition of the entries $\Gamma_{(x,a)}, \Gamma_{(y,b)}$ and $\Gamma_{(x,a)(y,b)}$ of Γ. The variables of the optimization are $\lambda \in \mathbb{R}$ and the other entries of Γ. These can also be taken real without loss of generality, because the constrained entries are real. Indeed, suppose that one finds possibly complex $\Gamma \geq 0$ which satisfies the constraints: Then its complex conjugate Γ_{sol}^* satisfies the constraints too; and therefore, the real part $\Gamma = \frac{1}{2}(\Gamma_{\text{sol}} + \Gamma_{\text{sol}}^*)$ satisfies the constraints as well.

If the algorithm yields $\Lambda < 0$, it is impossible to find $\Gamma \geq 0$ that satisfies the constraints: Then certainly $\mathcal{P} \notin Q$. If $\Lambda \geq 0$, $\mathcal{P} \in Q_{\mathcal{F}} \supset Q$, so we are not sure that $\mathcal{P} \in Q$. The assessment can be tightened by choosing a larger collection $\mathcal{F}$. But the only conclusive way to prove that $\mathcal{P} \in Q$ is to exhibit an explicit realization of the behavior in terms of state and projectors (having which, the SDP becomes superfluous).

6.3.3   Example 2: The TLM bound

Suppose that the verifier has performed a (2, 2; 2, 2) Bell test and estimated the correlators $(E_{00}, E_{01}, E_{10}, E_{11})$. For these correlators to belong the quantum set, there must exist a state and projectors such that

$$E_{xy} = \text{Tr}\big[\rho(\Pi^x_{a=+1} - \Pi^x_{a=-1})(\Pi^y_{b=+1} - \Pi^y_{b=-1})\big] \equiv \text{Tr}(\rho F^A_x F^B_y). \qquad (6.11)$$

For the collection of operators $\mathcal{F} = (F^A_0, F^A_1, F^B_0, F^B_1)$ we have

$$\Gamma = \begin{pmatrix} 1 & u_1 & E_{00} & E_{01} \\ u_1 & 1 & E_{10} & E_{11} \\ E_{00} & E_{10} & 1 & u_2 \\ E_{01} & E_{11} & u_2 & 1 \end{pmatrix}. \qquad (6.12)$$

The diagonal elements are 1 since $F^2_j = \mathbb{I}$, due to the orthogonality conditions $\Pi^x_a \Pi^x_{a'} = \delta_{a,a'} \Pi^x_a$ and the same for the Π^y_b. Once again, it is not restrictive to assume $u_1, u_2 \in \mathbb{R}$ since all the constrained entries are real. The condition for $\Gamma \geq 0$ reads $\text{Arcsin}(E_{00}) + \text{Arcsin}(E_{01}) + \text{Arcsin}(E_{10}) - \text{Arcsin}(E_{11}) \leq \pi$, as can be proved in two different ways (Landau, 1988; Navascués et al., 2008). Thus the TLM condition (6.5) is an upper bound to the set of quantum correlations. Since we know it can be reached, it is right on the boundary.

6.3.4   Example 3: Maximal quantum violation of a Bell inequality

Finally, let us consider the problem of finding the maximal violation of a Bell inequality in quantum theory, which was actually the first application of SDPs in the context of nonlocality (Wehner, 2006).

Assuming that the inequality reads $I \leq I_L$, the problem we want to solve is to compute

$$I_Q = \max I(\mathcal{P})$$
$$\text{subject to}\quad \mathcal{P} \in \mathcal{Q}. \qquad (6.13)$$

Unless one finds analytical arguments like the Tsirelson bound, this problem cannot be solved efficiently: At best, one can resort to heuristic optimization over a class of states and measurements, i.e., over a subset of $\mathcal{Q}$. Because of this, or because the optimization itself may have converged to a local maximum, the outcome satisfies $I_{\text{heur}} \leq I_Q$.

The NPA techniques allow relaxing the program (6.13) to the SDP

$$I_\Gamma = \max I(\mathcal{P})$$
$$\text{subject to}\quad \Gamma(\mathcal{P}) \geq 0[\text{i.e., } \mathcal{P} \in \mathcal{Q}_\mathcal{F}]. \qquad (6.14)$$

In this SDP, all the entries of Γ are variables. In order to obtain a non-trivial solution, all the $P(a, b|x, y)$ that appear in $I(\mathcal{P})$ must correspond to an entry in Γ: This defines the minimal collection $\mathcal{F}$. The SDP outputs both an upper and a lower bound on I_Γ, that usually coincide within numerical precision. Now we are guaranteed to have $I_\Gamma \geq I_Q$, because we are optimizing over a superset of $\mathcal{Q}$. The larger the collection of operators

$\mathcal{F}$, the closer I_Γ will be to the actual I_Q. Eventually, if I_Γ is found equal to an explicit achievable quantum value (a numerical I_{heur} or the result of some calculation), then we have found I_Q. This is how, in chapters 4 and 5, we could confidently cite the quantum maximum of several inequalities.

· ·

EXERCISES

These exercises involve running SDPs on a computer. Packages can be found online for all the major mathematical softwares.

Exercise 6.1 *As we shall see in section 7.1, the only quantum behavior that achieves the Tsirelson bound $S = 2\sqrt{2}$ is the behavior (3.11), characterized by $\langle A_x \rangle = \langle B_y \rangle = 0$ and $\langle A_x B_y \rangle = (-1)^{xy} \frac{1}{\sqrt{2}}$. In this exercise, we propose to rediscover this result by running SDPs. For instance, to prove that $\langle A_x \rangle = 0$ is the only solution, you need to run two SDPs, those with objectives $\max \langle A_x \rangle$ and $\min \langle A_x \rangle$.*

    (a) *Run your checks at the level $\mathcal{Q}_1$ of the hierarchy, i.e., for a moment matrix built on $\mathcal{F} = \{1, A_0, A_1, B_0, B_1\}$. Keep all the entries of the moment matrix as variables (apart from the entry that is 1) and impose only the constraint $S = 2\sqrt{2}$. You should find that this level of the hierarchy is not enough to certify uniqueness.*

    (b) *Repeat for the level $\mathcal{Q}_{1+AB}$ of the hierarchy, i.e., for a moment matrix built on $\mathcal{F} = \{1, A_0, A_1, B_0, B_1, A_0 B_0, A_0 B_1, A_1 B_0, A_1 B_1\}$. This time, you should be able to prove uniqueness (up to numerical precision).*

Exercise 6.2. *Using the NPA method at the level $\mathcal{Q}_1$, prove that the quantum minimum for Mermin's outreach Bell test (subsections 1.4.1 and 4.2.2) is $O = 4.5$.*

7
Device-Independent Self-Testing

You can observe a lot by just watching.

Yogi Berra

Our first contact with DI certification is device-independent self-testing, that was mentioned a few times in chapters 3–5: In a sense to be made precise in this chapter, some behaviors are possible only with a unique choice of state and measurements.

7.1 Self-Testing of the Maximal Violation of CHSH

As a first case study, we are going to show that $S=2\sqrt{2}$ for the CHSH test can only be reached by measuring a two-qubit maximally entangled state with projective measurements in mutually unbiased bases. We follow (Popescu and Rohrlich, 1992*b*); the same result can be identified in previous works, in a more abstract mathematical framework (Tsirel'son, 1987; Summers and Werner, 1987).

7.1.1 The proof

We resume from the proof of the Tsirelson bound (subsection 3.2.2), working with projective measurements without loss of generality since the dimension of the Hilbert space is not fixed. In this case, the operators A_0, A_1, B_0 and B_1 have eigenvalues -1 and $+1$; and we saw that $S = 2\sqrt{2}$ is only achievable if

$$\mathrm{Tr}\,(\rho\,[A_0, A_1] \otimes [B_0, B_1]) = -4. \tag{7.1}$$

Given that $||[A_0, A_1]||_\infty \leq 2$ and $||[B_0, B_1]||_\infty \leq 2$, this implies that $[A_0, A_1]$ and $[B_0, B_1]$ must have, in some subspaces, the eigenvalue $+2i$ or $-2i$. Let's denote those subspaces by their projectors $\Pi^{A,B}_{+,-}$. In all those subspaces, the anticommutators $\{A_0, A_1\}$ and $\{B_0, B_1\}$ have the eigenvalue 0, since $[A_0, A_1]^2 = \{A_0, A_1\}^2 - 4\mathbb{I}$ holds using $A_0^2 = A_1^2 = \mathbb{I}$.

Besides, the support of ρ must be in the right subspaces for (7.1) to hold:

$$\rho = \sum_{s=\pm}(\Pi^A_s \otimes \Pi^B_s)\rho(\Pi^A_s \otimes \Pi^B_s). \tag{7.2}$$

Bell Nonlocality. Valerio Scarani. © Valerio Scarani 2019. Published in 2019 by Oxford University Press.
DOI: 10.1093/oso/9780198788416.001.0001

To keep the notation simple, in what follows we restrict A_x and B_y to those subspaces, i.e., A_x actually stands for $\sum_{s=\pm} \Pi_s^A A_x \Pi_s^A$ and B_y for $\sum_{s=\pm} \Pi_s^B B_y \Pi_s^B$; the more general notation will be re-introduced in the next subsection.

Having noticed this, the following can be proved using Jordan's lemma (see Appendix G.4) or in other ways (Kaniewski, 2017): If there is a subspace of $\{A_0, A_1\}$ associated to the eigenvalue 0, it must be of even dimension $2d''_A$; besides, there exists a choice of basis in which

$$A_0 = \sigma_{\hat{z}} \otimes \mathbb{I}_{d''_A} \quad \text{and} \quad A_1 = \sigma_{\hat{x}} \otimes \mathbb{I}_{d''_A}. \tag{7.3}$$

The same of course holds on Bob's side. Let's now plug the expressions for the local operators for the measurements into the Bell operator (3.4). To simplify this step we choose Bob's basis such that $B_0 = \frac{1}{\sqrt{2}}(\sigma_{\hat{z}} + \sigma_{\hat{x}}) \otimes \mathbb{I}_{d''_B}$ and $B_1 = \frac{1}{\sqrt{2}}(\sigma_{\hat{z}} - \sigma_{\hat{x}}) \otimes \mathbb{I}_{d''_B}$. Then $\mathbf{S} = \sqrt{2}(\sigma_{\hat{z}} \otimes \sigma_{\hat{z}} + \sigma_{\hat{x}} \otimes \sigma_{\hat{x}}) \otimes \mathbb{I}_{d''_A} \otimes \mathbb{I}_{d''_B}$ and $\text{Tr}(\mathbf{S}\rho) = 2\sqrt{2}$ can be achieved if and only if $\langle \sigma_{\hat{z}} \otimes \sigma_{\hat{z}} \rangle = \langle \sigma_{\hat{x}} \otimes \sigma_{\hat{x}} \rangle = 1$. This requirement uniquely identifies the Bell state $|\Phi^+\rangle$. Therefore we conclude

$$\rho = |\Phi^+\rangle\langle\Phi^+| \otimes \tilde{\rho}, \tag{7.4}$$

where $\tilde{\rho}$ can be any state.

In words, we have established that, in order to achieve $S = 2\sqrt{2}$, Alice and Bob must share a two-qubit maximally entangled state and perform projective measurements in mutually unbiased bases. This is self-testing of both the state and the measurements.

7.1.2 Consolidation

Having proved the self-testing character of $S = 2\sqrt{2}$ in a series of inferences, let us summarize the result with a more precise notation. Consider the following decomposition of Alice and Bob's Hilbert spaces:

$$\mathcal{H}_A = [\mathcal{H}_{A'} \otimes \mathcal{H}_{A''}] \oplus \mathcal{H}_{A'''} \tag{7.5}$$

$$\mathcal{H}_B = [\mathcal{H}_{B'} \otimes \mathcal{H}_{B''}] \oplus \mathcal{H}_{B'''} \tag{7.6}$$

where $\dim(\mathcal{H}_K) \equiv d_K$ unspecified, $\dim(\mathcal{H}_{K'}) = 2$, $\dim(\mathcal{H}_{K''}) \equiv d''_K \leq d_K/2$ but otherwise unspecified, and $\dim(\mathcal{H}_{K'''}) = d_K - 2d''_K$, for $K = A, B$.

Then *up to local unitaries* the state reads

$$\rho_{AB} = |\Phi^+\rangle\langle\Phi^+| \otimes \tilde{\rho}_{A''B''} \tag{7.7}$$

with no support in $\mathcal{H}_{A'''}$ and $\mathcal{H}_{B'''}$. Notice that $\tilde{\rho}$ is arbitrary and could be entangled.

With the same choice of basis, the measurement operators read

$$\Pi_a^x = \left[\left(\frac{\mathbb{I} + a\hat{a}_x \cdot \vec{\sigma}}{2} \right)_{A'} \otimes \mathbb{I}_{A''} \right] \oplus \left(\tilde{\Pi}_a^x \right)_{A'''} \tag{7.8}$$

$$\Pi_b^y = \left[\left(\frac{\mathbb{I} + b\hat{b}_y \cdot \vec{\sigma}}{2} \right)_{B'} \otimes \mathbb{I}_{B''} \right] \oplus \left(\tilde{\Pi}_b^y \right)_{B'''} \tag{7.9}$$

where the $\tilde{\Pi}$ are arbitrary and where

$$\hat{a}_0 = \hat{z}, \ \hat{a}_1 = \hat{x}; \qquad \hat{b}_0 = \frac{\hat{z} + \hat{x}}{\sqrt{2}}, \ \hat{b}_1 = \frac{\hat{z} - \hat{x}}{\sqrt{2}}. \tag{7.10}$$

This is what is meant by saying that $S = 2\sqrt{2}$ *self-tests the maximally entangled state of two-qubits and mutually unbiased measurements on it by both players*. As a consequence of self-testing, there is only one quantum behavior that reaches $S = 2\sqrt{2}$, namely (3.11).

7.1.3   Two worked out examples

For pedagogical reasons let us consider two states that lead to $S = 2\sqrt{2}$ and show explicitly that they can indeed be written in the form (7.7).

The first example is a state of the form

$$|\Psi\rangle = \sum_{j=1}^{d/2} \sqrt{p_j} \frac{|2j-1\rangle|2j-1\rangle + |2j\rangle|2j\rangle}{\sqrt{2}} \tag{7.11}$$

with d even. The fact that $S = 2\sqrt{2}$ can be achieved with this state is a particular case of the calculation done in section 3.2.4 with the state (3.20). By recording the parity of the ket label in a qubit and the value of j in a $(d/2)$-dimensional subsystem, i.e., $|2j-1\rangle_K \equiv |1\rangle_{K'} \otimes |j\rangle_{K''}$ and $|2j\rangle_K \equiv |0\rangle_{K'} \otimes |j\rangle_{K''}$ for $K = A, B$, the state is indeed cast in the form (7.7)

$$|\Psi\rangle = \left(|\Phi^+\rangle \right)_{A'B'} \otimes \left(\sum_{j=1}^{d/2} \sqrt{p_j} |j\rangle_{A''} |j\rangle_{B''} \right). \tag{7.12}$$

We started from a pure state (7.11) that is a superposition of singlets in different subspaces; but the same isometry could have extracted a singlet in $A'B'$ even if we had started from an incoherent mixture of those singlets. Thus, one can violate CHSH maximally with a mixed state, as first observed by (Braunstein *et al.*, 1992). A similar case study is proposed as Exercise 7.1.

As a second example, we consider the *Smolin state* of four qubits

$$\rho_{ABCD} = \frac{1}{4} \sum_{k=1}^{4} \left(\Pi_{\Phi^+\Phi^+} + \Pi_{\Phi^-\Phi^-} + \Pi_{\Psi^+\Psi^+} + \Pi_{\Psi^-\Psi^-} \right) \tag{7.13}$$

with[1] $\Pi_{\psi\psi} \equiv (|\psi\rangle\langle\psi|)_{AB}(|\psi\rangle\langle\psi|)_{CD}$. If the four players can only apply local operations and classical communications, this is a "bound entangled" state. However, it was observed that if $BCD \equiv E$ are treated as a single player holding a 8-dimensional system, one can reach $S = 2\sqrt{2}$ with measurements on this state (Augusiak and Horodecki, 2006).

In order to bring the Smolin state to the form (7.7), we recall that $|\Phi^+\rangle_{AB} = \mathbb{I} \otimes \sigma_{\hat{z}}|\Phi^-\rangle_{AB} = \mathbb{I} \otimes \sigma_{\hat{x}}|\Psi^+\rangle_{AB} = \mathbb{I} \otimes (i\sigma_{\hat{y}})|\Psi^-\rangle_{AB}$. Therefore, E can apply the unitary

$$U_E = \mathbb{I} \otimes |\Phi^+\rangle\langle\Phi^+| + \sigma_{\hat{z}} \otimes |\Phi^-\rangle\langle\Phi^-|$$

$$+ \sigma_{\hat{x}} \otimes |\Psi^+\rangle\langle\Psi^+| + (i\sigma_{\hat{y}}) \otimes |\Psi^-\rangle\langle\Psi^-| \tag{7.14}$$

where the Bell basis is on CD. Then

$$U_E \rho_{ABCD} U_E^\dagger = \left(|\Phi^+\rangle\langle\Phi^+|\right)_{AB} \otimes \left(\frac{1}{4}\mathbb{I}\right)_{CD}. \tag{7.15}$$

7.2   The Mayers-Yao Self-Testing Behavior

Our second case study follows the work of Mayers and Yao (1998; 2004) in the context of quantum cryptography. The original work was in the (3, 2; 3, 2) Bell scenario, but we present a variation in the (2, 2; 3, 2) scenario (McKague *et al.*, 2012). The outputs are labeled $a, b \in \{-1, +1\}$, so we can parametrize

$$P(a, b|x, y) = \frac{1}{4}\left(1 + a\langle A_x\rangle + b\langle B_y\rangle + ab\langle A_x B_y\rangle\right).$$

The *MY behavior* is defined by

$$\langle A_x\rangle = \langle B_y\rangle = 0, \quad x \in \{0, 1\}, \quad y \in \{0, 1, 2\} \tag{7.16}$$

$$\langle A_0 B_0\rangle = \langle A_1 B_1\rangle = 1 \tag{7.17}$$

$$\langle A_0 B_1\rangle = \langle A_1 B_0\rangle = 0 \tag{7.18}$$

$$\langle A_0 B_2\rangle = \langle A_1 B_2\rangle = \frac{1}{\sqrt{2}}. \tag{7.19}$$

---

[1] Notice that ρ remains the same if we were to use $\Pi_{\psi\psi} \equiv (|\psi\rangle\langle\psi|)_{AC}(|\psi\rangle\langle\psi|)_{BD}$ or $\Pi_{\psi\psi} \equiv (|\psi\rangle\langle\psi|)_{AD}(|\psi\rangle\langle\psi|)_{BC}$.

This is the behavior that one obtains by measuring the state $\left|\Phi^+\right\rangle$ with $A_0 = B_0 = \sigma_{\hat{z}}$, $A_1 = B_1 = \sigma_{\hat{x}}$ and $B_2 = \frac{1}{\sqrt{2}}(\sigma_{\hat{z}} + \sigma_{\hat{x}})$. We need to prove that this is actually the only possible realization, in the sense of self-testing.

As a sanity check, notice that the MY behavior is nonlocal:[2] It violates CHSH as $\langle A_0 B_2\rangle + \langle A_0 B_0\rangle + \langle A_1 B_2\rangle - \langle A_1 B_0\rangle = \sqrt{2} + 1 > 2$. It's also easy to check by inspection that this is the highest violation of CHSH achievable by combining the statistics in the behavior: Since $\sqrt{2} + 1 < 2\sqrt{2}$, this form of self-testing is genuinely different from that of the previous section.

7.2.1   Inferences from the behavior

Just as in the previous section, we work with projective measurements without loss of generality. Also, here the proof is more transparent by working with a pure state (without loss of generality, as explained in subsection 6.2.1).

Thus we read the $\langle A_x B_y\rangle$ as $\langle \Psi | A_x B_y | \Psi \rangle$. The condition (7.17) can be rewritten as

$$A_0 |\Psi\rangle = B_0 |\Psi\rangle \quad \text{and} \quad A_1 |\Psi\rangle = B_1 |\Psi\rangle. \tag{7.20}$$

Inserting this into (7.18), we find $\langle \Psi | A_0 A_1 | \Psi \rangle = 0$, that is $A_1 |\Psi\rangle$ is orthogonal to $A_0 |\Psi\rangle$. If this is the case, then (7.19) means that

$$B_2 |\Psi\rangle = \frac{A_0 + A_1}{\sqrt{2}} |\Psi\rangle: \tag{7.21}$$

indeed, $B_2 |\Psi\rangle$ must be of norm 1, and we know already two of its projections of amplitude $\frac{1}{\sqrt{2}}$ on two orthogonal vectors. The last preparatory step consists in computing $B_2^2 |\Psi\rangle$. On the one hand, we know that $B_2^2 = \mathbb{I}_{d_B}$; on the other hand, using $[A_x, B_y] = 0$ we obtain[3]

$$B_2^2 |\Psi\rangle = B_2 \frac{A_0 + A_1}{\sqrt{2}} |\Psi\rangle = \frac{A_0 + A_1}{\sqrt{2}} B_2 |\Psi\rangle = \left(\frac{A_0 + A_1}{\sqrt{2}}\right)^2 |\Psi\rangle$$

$$= \frac{1}{2}(A_0^2 + A_1^2 + \{A_0, A_1\}) |\Psi\rangle.$$

---

[2] Without Bob's third input $y = 2$, the behavior becomes local. That behavior self-tests the singlet if one assumes *a priori* the systems to be qubits (Appendix F.1.3).

[3] Notice that, from an equation of the form $C|\Psi\rangle = D|\Psi\rangle$ like (7.21), the inference $C^2 |\Psi\rangle = D^2 |\Psi\rangle$ follows *automatically* only if $C|\Psi\rangle \propto |\Psi\rangle$, which we cannot guarantee in a device-independent setting (and in fact, it would be wrong if one replaces the ideal states and operators). This is why we have to *prove* that $C^2 |\Psi\rangle = D^2 |\Psi\rangle$ holds here, by using the commutation relations.

Since $A_0^2 = A_1^2 = \mathbb{I}_{d_A}$, we find finally

$$\{A_0, A_1\}|\Psi\rangle = 0 \overset{7.20}{\Longrightarrow} \{B_0, B_1\}|\Psi\rangle = 0. \tag{7.22}$$

We can now concentrate on inputs x, $y = 0$, 1, Bob's third measurement $y = 2$ has fulfilled its role. In fact (7.22) are the same conditions as we found in the previous section by reasoning on the algebra behind $S = 2\sqrt{2}$. At this point, we could finish the proof by following the same pattern: Invoke Jordan's lemma and solve an easy two-qubit problem. Instead, we take a more complicated approach, which however can lead to generalizations—in fact, it anticipates the formal definition of self-testing that will be given in subsection 7.3.1.

7.2.2   Swapping the relevant qubit into a controlled one

Assume self-testing indeed holds: On Alice's side there is then one qubit that is maximally entangled with one on Bob's side, and the operators A_0 and A_1 will act non-trivially only on the qubit support of the state. We can now imagine the following virtual protocol: Alice prepares a controlled qubit A' in a dummy state, say $|0\rangle$, then couples it to her system in the attempt of *swapping* the state of the relevant qubit and that of the controlled qubit. If Bob does the same and the swap is correctly implemented, at the end of these local operations we expect to find the state $|\Phi^+\rangle_{A'B'}$ in the controlled qubits.

A possible construction of the operator that swaps the state of two qubits is shown in Figure 7.1, left. For the system under study, we do not have the Pauli operators, but

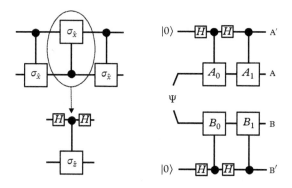

Figure 7.1 *Left: Quantum circuit representation of the swap gate for two qubits. Left, top: The usual representations as three alternate CNOT gates: For a given control c and target t, the gate is defined as $U_{ct} = |0\rangle\langle 0| \otimes \mathbb{I} + |1\rangle\langle 1| \otimes \sigma_{\hat{x}}$, i.e., the gate acts non-trivially only if the control is in the state $|1\rangle$. Left, bottom: An equivalent version of the central CNOT gate, where H is the Hadamard unitary gate defined as usual: $H|0\rangle = \frac{1}{\sqrt{2}}(|0\rangle + |1\rangle), H|1\rangle = \frac{1}{\sqrt{2}}(|0\rangle - |1\rangle)$). Right: The corresponding isometry in the context of Mayers-Yao self-testing. Having chosen the auxiliary qubits A' and B' in the state $|0\rangle$, the first controlled gate will act trivially and can therefore be dispensed with. In the remaining two controlled gates, $\sigma_{\hat{z}}$ is replaced by A_0 (B_0) and $\sigma_{\hat{x}}$ is replaced by A_1 (B_1).*

we surmise that A_0 and B_0 act as $\sigma_{\hat{z}}$, A_1 and B_1 as $\sigma_{\hat{x}}$, on the relevant qubits. So the virtual protocol is implemented as the *local isometry* $\Phi = \Phi_A \otimes \Phi_B$ represented in Figure 7.1, right. Finally, let us show that the virtual protocol just defined implements indeed the swap of the relevant qubits if (7.20) and (7.22) hold. Omitting the tensor product symbol for simplicity:

$$\Phi |\Psi\rangle_{AB} |00\rangle_{A'B'} = \frac{1}{4} \left[[(\mathbb{I}+A_0)(\mathbb{I}+B_0) |\Psi\rangle] |00\rangle \right.$$
$$+ [A_1 B_1 (\mathbb{I}-A_0)(\mathbb{I}-B_0) |\Psi\rangle] |11\rangle$$
$$+ [B_1 (\mathbb{I}+A_0)(\mathbb{I}-B_0) |\Psi\rangle] |01\rangle$$
$$\left. + [A_1 (\mathbb{I}-A_0)(\mathbb{I}+B_0) |\Psi\rangle] |10\rangle \right]. \tag{7.23}$$

First, using (7.20), we replace B_0 with A_0, and this cancels the third and fourth lines because $(\mathbb{I}+A_0)(\mathbb{I}-A_0) = \mathbb{I}-A_0^2 = 0$. Then, using (7.22), one proves that $A_1 B_1 (\mathbb{I}-A_0)(\mathbb{I}-B_0) |\Psi\rangle = (\mathbb{I}+A_0)(\mathbb{I}+B_0)A_1 B_1 |\Psi\rangle$, which is in turn equal to $(\mathbb{I}+A_0)(\mathbb{I}+B_0) |\Psi\rangle$ because of (7.20). Finally $(\mathbb{I}+A_0)(\mathbb{I}+B_0) |\Psi\rangle = (\mathbb{I}+A_0)^2 |\Psi\rangle = 2(\mathbb{I}+A_0) |\Psi\rangle$. So we have found

$$\Phi |\Psi\rangle_{AB} |00\rangle_{A'B'} = \left(\frac{\mathbb{I}+A_0}{\sqrt{2}} |\Psi\rangle \right)_{AB} |\Phi^+\rangle_{A'B'} \tag{7.24}$$

which is the self-testing of the state.

The proof for the measurements follows the same steps, starting with $\Phi A_x B_y |\Psi\rangle_{AB} |00\rangle_{A'B'}$ instead of (7.23). Let me show it for one of the six cases (for clarity, I underline the operator under study):

$$\Phi \underline{A_1} |\Psi\rangle_{AB} |00\rangle_{A'B'} = \frac{1}{4} \left[[(\mathbb{I}+A_0)(\mathbb{I}+B_0)\underline{A_1} |\Psi\rangle] |00\rangle \right.$$
$$+ [A_1 B_1 (\mathbb{I}-A_0)(\mathbb{I}-B_0)\underline{A_1} |\Psi\rangle] |11\rangle$$
$$+ [B_1 (\mathbb{I}+A_0)(\mathbb{I}-B_0)\underline{A_1} |\Psi\rangle] |01\rangle$$
$$\left. + [A_1 (\mathbb{I}-A_0)(\mathbb{I}+B_0)\underline{A_1} |\Psi\rangle] |10\rangle \right].$$

By using (7.22), one moves A_1 to the left in the first, second, and fourth lines while changing the sign of A_0, and B_1 to the right in the third while changing the sign of B_0. After simplifications similar to those made previously, the final result is

$$\Phi \underline{A_1} |\Psi\rangle_{AB} |00\rangle_{A'B'} = \left(\frac{\mathbb{I}+A_0}{\sqrt{2}} |\Psi\rangle \right)_{AB} \left(\underline{\sigma_{\hat{x}} \otimes \mathbb{I} |\Phi^+\rangle} \right)_{A'B'}. \tag{7.25}$$

Notice that the state left in the AB system is the same for both the self-testing of the state and of the measurements.

7.3 Formal Definition and its Consequences

7.3.1 Formal definition of self-testing

Having gained confidence with self-testing by dealing with two examples, we can now state the formal *definition of self-testing*. We write it for bipartite Bell scenarios, the extension to multipartite scenarios is obvious.

The behavior $\mathcal{P}$ self-tests the state $|\bar{\psi}\rangle$ and the families of measurements $\{\bar{\Pi}_a^x, x \in \mathcal{X}, a \in \mathcal{A}\}$ and $\{\bar{\Pi}_b^y, y \in \mathcal{Y}, b \in \mathcal{B}\}$ if the following holds: For every quantum realization $P(a,b|x,y) = \mathrm{Tr}(\rho \Pi_a^x \Pi_b^y)$ of the behavior, there exist a local isometry $\Phi = \Phi_A \otimes \Phi_B$ such that

$$\Phi[\rho] = |\bar{\psi}\rangle\langle\bar{\psi}|_{A'B'} \otimes \tilde{\rho}_{A''B''} \tag{7.26}$$

$$\Phi_A[\Pi_a^x] = \left[\left(\bar{\Pi}_a^x \right)_{A'} \otimes \mathbb{I}_{A''} \right] \oplus \left(\tilde{\Pi}_a^x \right)_{A'''} \tag{7.27}$$

$$\Phi_B[\Pi_b^y] = \left[\left(\bar{\Pi}_b^y \right)_{B'} \otimes \mathbb{I}_{B''} \right] \oplus \left(\tilde{\Pi}_b^y \right)_{B'''}. \tag{7.28}$$

where the tilde operators are arbitrary. The isometries have been defined as $\Phi_K : \mathcal{H}_K \to (\mathcal{H}_{K'} \otimes \mathcal{H}_{K''}) \oplus \mathcal{H}_{K'''}$, $K = A, B$, where the dimension of $\mathcal{H}_{K'}$ is fixed by the state and measurement that is self-tested, while the dimensions of $\mathcal{H}_{K''}$ and $\mathcal{H}_{K'''}$ are arbitrary.

Even if it should be clear, let me stress that an experimental demonstration of self-testing does not require setting up the local isometries. It is sufficient to perform a normal Bell test and record the behavior, all the rest are mathematical inferences drawn from that observation. Let me also insert two technical remarks:

- The two examples of proofs that we presented differ in one respect: In section 7.1 K' and K'' appeared as subsystems of the original K, whereas in section 7.2 the K' are auxiliary systems and the K'' are the original systems K. Only in the former case self-testing can be described as "swapping out of the box" the relevant degrees of freedom. For the formal definition of self-testing, we retained the more general definition of isometry, which is the second.[4]

- Given a quantum realization of a behavior, another realization related by *any* isometry will give the same behavior, not just a local one. Besides, if one were to identify subsystems with algebras of commuting observables (see subsection 6.2.2), it would be hard to define "local" isometries, because $[A, B] = 0$ implies $[\Phi[A], \Phi[B]] = 0$ for any isometry. In fact, if the goal is to define the set of behaviors

---

[4] The most striking example of the power of auxiliary systems is a very counter-intuitive result in quantum channel theory. One would expect an ancilla prepared in the maximally mixed state to describe a source of classical randomness; thus, channels that use such an ancilla should be equivalent to mixtures of unitary operations on the system. But this is not the case: There exist unital channels that cannot be represented as convex mixtures of unitary operations (Haagerup and Musat, 2011).

that lead to self-testing, the locality requirement can be dropped from the isometry. However, locality must be enforced if self-testing is seen as DI certification of given black boxes, because under global isometries the state that gives the behavior could be in the auxiliary systems. Besides, if not based on locality, the assessment of $[A, B] = 0$ requires characterization of the devices.

7.3.2   Which behaviors, and which states

Next, we are going to see which behaviors may lead to self-testing, and which states can be the object of self-testing [see Šupić and Bowles, 2019) for a more comprehensive review].

It was proved that only behaviors that are *extremal points* of Q may lead to self-testing (Goh *et al.*, 2018). One may conjecture that *all* the extremal points of Q in any Bell scenario self-test some state, but there is no proof for this. Recall that several behaviors may self-test the same state, as we have seen the two-qubit maximally entangled state being self-tested by $S = 2\sqrt{2}$, by the Mayers-Yao correlations, and in fact by all the nonlocal points on the TLM boundary (6.5) (Tsirel'son, 1987; Wang *et al.*, 2016).

It follows from this that *mixed state* cannot be self-tested: Behaviors obtained from them are not extremal in the quantum set, since they can be written as the convex mixture of behaviors obtained from pure states with the same measurements. This observation motivates *a posteriori* our defining self-testing with pure states.[5]

It is known that *all pure bipartite entangled states can be self-tested* (Coladangelo *et al.*, 2017a). Several examples of self-testable *multipartite pure entangled states* are known too: The GHZ states and in fact all graph states, the three-qubit $|W\rangle$ state and several of its generalizations, and others [see (Šupić *et al.*, 2018) and references therein]. For generic multipartite states, however, self-testing may have to be defined in a more careful way, as there exist states that are not equivalent under local unitaries but whose state-behavior is identical.[6]

7.4   Approximate Self-Testing: Robustness Bounds

As we have just seen, self-testing *per se* is a property of extremal behaviors and pure states. These are theoretical ideal cases. For self-testing to enter the realm of device certification, one must be able to make statements on any observed behavior. For instance, if we observe $S = 2\sqrt{2} - \varepsilon$, we should be able to estimate *how close* the actual state is to the

---

[5] Besides, a behavior that can be obtained with a mixed state can also be obtained with its purification, using measurements that act trivially on the purifying degrees of freedom. Thus, as a corollary, we know that if a behavior self-tests a pure state $|\bar{\psi}\rangle$, the systems A' and B' cannot be further decomposed in sub-systems, such that the measurements would act trivially on one of them.

[6] If a behavior can be obtained with state $|\psi\rangle$ and some measurements, that same behavior can also be obtained with the state $|\psi\rangle^*$, defined as taking the complex conjugation of the coefficients in some basis, and the similarly conjugated measurements. For bipartite systems, $|\psi\rangle$ and $|\psi\rangle^*$ are always equivalent under local unitaries, the canonical form being the Schmidt decomposition whose coefficients are real and positive. However, this equivalence no longer holds for any systems composed of three or more subsystems (Acín *et al.*, 2001). For a solution in network setting, see (Šupić et al. 2023).

ideal one. Such an estimate is called a *robustness bound*. Let us review quickly some of the results, referring to the original works for more details. It must be kept in mind that all robustness calculations rely on *choosing a form of the local isometry* Φ with which to do the calculation. By default, one chooses the isometry of Figure 7.1, even if it won't be a swap in the non-ideal case and other isometries may be found that give better bounds (Bancal *et al.*, 2015).

7.4.1 Generic analytical bounds with poor robustness

The figure of merit for robustness was initially defined as (McKague *et al.*, 2012; Reichardt *et al.*, 2013)

$$\Delta = ||\Phi|\Psi\rangle - |\bar{\psi}\rangle_{A'B'} \otimes |\tilde{\psi}\rangle_{A''B''}|| \tag{7.29}$$

where the unknown state is supposed to be pure. The calculation amounts at enchaining inequalities (triangle, Cauchy-Schwarz …) and can be applied in principle to any example of self-testing; the resulting bounds are analytical expressions. The problem of these bounds is their very poor tolerance. For instance, for $S = 2\sqrt{2} - \varepsilon$, the estimate of (McKague *et al.*, 2012) yields

$$\Delta \leq 11 \left(\varepsilon\sqrt{2}\right)^{1/2} + 10 \left(\varepsilon\sqrt{2}\right)^{1/4}. \tag{7.30}$$

As a benchmark, let us consider the situation in which the isometry does not couple the auxiliary systems to the unknown system, that is if $\Phi|\Psi\rangle = |00\rangle_{A'B'} \otimes |\tilde{\psi}\rangle_{A''B''}$. For this state one finds $\Delta = \sqrt{2 - \sqrt{2}}$, which is reached already for $\varepsilon \approx 1.8 \times 10^{-5}$. For higher values of ε, this criterion cannot guarantee anything non-trivial. While this is sufficient to establish robustness in principle, much better tolerance is needed to make meaningful self-testing claims on experimental data.

7.4.2 Generic SDP techniques

The desired better tolerance has been obtained by resorting to SDP optimizations based on the NPA hierarchy. Here, the figure of merit should be linear in the behavior in order to play the role of objective function in the SDP. The natural choice is to *minimize the fidelity* $F = \langle \bar{\psi} | \rho_{A'B'} | \bar{\psi} \rangle$ of the reduced state of the auxiliary systems with the ideal output state. Given the expression of F, the SDP needs to find the minimal value of F compatible with the observed behavior and for some choice of the moment matrix Γ. The fidelity F is a sum of terms, some of which may be determined by the behavior, while the others must appear as semi-definite variables in the moment matrix. Still, the choice of Γ adds to the choice of the isometry in making the bound possibly not tight.

For the sake of an example, let us consider the Mayers-Yao criterion. We can find the expression of the fidelity starting from (7.23):

$$F = \left\| \frac{1}{4\sqrt{2}} \left[(\mathbb{I}+A_0)(\mathbb{I}+B_0) + A_1 B_1 (\mathbb{I}-A_0)(\mathbb{I}-B_0) \right] |\Psi\rangle \right\|^2$$
$$= \frac{1}{16} \left[4 + 4\langle A_0 B_0 \rangle + \langle A_1 B_1 \rangle + \langle [A_0, A_1][B_0, B_1] \rangle \right.$$
$$\left. - \langle A_1 (B_0 B_1 B_0) \rangle - \langle (A_0 A_1 A_0) B_1 \rangle + \langle (A_0 A_1 A_0)(B_0 B_1 B_0) \rangle \right] \qquad (7.31)$$

where $\langle C \rangle = \langle \Psi | C | \Psi \rangle$. The values of the terms $\langle A_0 B_0 \rangle$ and $\langle A_1 B_1 \rangle$ are determined by the behavior; the other terms are variables. For the CHSH criterion, the expression of the fidelity is different (Exercise 7.2). If in the SDP the only constraint coming from the behavior is $S = 2\sqrt{2} - \varepsilon$, the benchmark value $F = \frac{1}{2}$ compatible with the output state $|00\rangle_{A'B'}$ is reached for $\varepsilon \approx 0.4$. This tolerance may be improved by using the whole behavior as constraint, as well as by changing the isometry (Bancal *et al.*, 2015).

This approach through SDPs can also be applied in principle to any example of self-testing. However, moving away from the basic examples, the number of terms in the expression of F grows quickly, and so does the SDP size. Besides, if the natural operators that define the correlations are not those that define a swap and/or do not guarantee unitarity, additional semi-definite constraints must be used. In practice, the self-testing of the maximal violation of the CGLMP inequality for $m = 3$ (Bancal *et al.*, 2015), or the self-testing of two singlets using the Magic Square game (Wu *et al.*, 2016), are already very cumbersome and computationally heavy.

7.4.3   A highly robust analytical result for specific cases

Analytical bounds with even better tolerance than the SDP method have been obtained more recently for Bell scenarios with two inputs and two outputs per player (Kaniewski, 2016). For the bipartite case, which is CHSH, the lower bound on the fidelity is given by

$$F(S) = \frac{1}{2} + \frac{1}{2}\frac{S - S^*}{2\sqrt{2} - S^*}, \quad \text{with } S^* = \frac{16 + 14\sqrt{2}}{17} \approx 2.11. \qquad (7.32)$$

In other words, the value $F = \frac{1}{2}$ is reached for $\varepsilon = 2\sqrt{2} - S^* \approx 0.72$. At the moment of writing, we do not know if this bound is tight, that is, if there is a family of states that saturates it (Exercise 7.3).

. . . .

EXERCISES

Exercise 7.1. *The "linear cluster state" of N qubits is a pure state that is not product in any partition; in particular, all n-qubit partial states are mixed unless n = N. For N > 5, the partial state of any five consecutive qubits is such that the following expectation values hold:*

$$\langle IZXZI \rangle = \langle ZYYZI \rangle = \langle IZYYZ \rangle = -\langle ZYXYZ \rangle = +1 \qquad (7.33)$$

where $I \equiv \mathbb{I}$, $Z \equiv \sigma_{\hat{z}}$ *and* $X \equiv \sigma_{\hat{x}}$.

    (a) *Prove that the set of equations (7.33) defines a GHZ-type argument for nonlocality: Thus, one can have GHZ arguments for mixed states (Scarani et al., 2005).*

    (b) *With a suitable grouping of some of the players, this becomes the usual 3-partite GHZ argument, which is known to self-test the state* $|GHZ_3\rangle$. *Prove that, under local isometries for that grouping, the equations (7.33) are equivalent to the set of eigenvalue equations that define that state (5.14).*

Exercise 7.2. *Prove that, for the usual choice of isometry (Figure 7.1), the singlet fidelity for self-testing using CHSH is given by*

$$F = \left\| \frac{1}{4\sqrt{2}} \left[(c+s)(\mathbb{I}+A_0)(\mathbb{I}+B_0) + (c-s)A_1 B_1 (\mathbb{I}-A_0)(\mathbb{I}-B_0) \right] |\Psi\rangle \right\|^2 \qquad (7.34)$$

with $c = \cos \frac{\pi}{8}$ *and* $s = \sin \frac{\pi}{8}$. *In particular, it is different from the Mayers-Yao expression (7.31). Hint: The isometry assumes that* $A_0 = B_0 = \sigma_{\hat{z}}$ *and* $A_1 = B_1 = \sigma_{\hat{x}}$ *in the ideal case. For this choice of operators, the state that gives* $S = 2\sqrt{2}$ *is* $c |\Phi^+\rangle + s |\Phi^-\rangle$.

Exercise 7.3. *Suppose that the state being measured is a two-qubit Werner state (3.23). Compare the singlet fidelity that one would obtain after characterizing the state with the DI self-testing bound (7.32) computed assuming that the observed S is the maximal value of CHSH achievable with that state.*

8

Certifying Randomness

Esse igitur contra rationem providentiae et perfectionis rerum si non essent aliqua casualia.

So, it would be contrary to providence and to the perfection of things if there were no chance events.

Thomas Aquinas, *Summa contra Gentiles*

If nonlocality is observed, the outputs of the process did not pre-exist—in other words, the process is random. In this respect, the quantification of the amount of generated randomness is the most natural example of device-independent certification. This chapter is devoted to it.

8.1 Introduction to Randomness

The word "randomness" evokes notions that span across metaphysics, anthropology, history, mathematics, and computer science: Chance, free will, gambling, secrecy, statistics, zero-knowledge proofs ... Delving into these cultural riches is exciting, but would distract us from the topic of this book. Our much more modest goal of this chapter is to discuss *the role of nonlocality in certifying randomness*. Even in this restrictive setting, a few concepts must be introduced. For two alternative introductions to the same topic, the reader can consult the review articles (Pivoluska and Plesch, 2014; Acín and Masanes, 2016).

8.1.1 Product and process randomness

First of all, randomness is studied in the context of *processes that produce strings of numbers*. One speaks both of *random processes* and of *random numbers*. The corresponding notions of randomness are called *process randomness* and *product randomness* (the randomness of the "product," that is, of the string of numbers).

We expect these two notions to be tightly connected: Random processes are often called *random number generators (RNGs)*. Nonetheless, they are different, and there are situations in which the difference is clear. On the one hand, a random process may occasionally output a number that nobody would call random, for instance a string

Bell Nonlocality. Valerio Scarani. © Valerio Scarani 2019. Published in 2019 by Oxford University Press.
DOI: 10.1093/oso/9780198788416.001.0001

of zeros. On the other hand, reading pre-recorded information would not be called a random process, even if the pre-recorded information is a random number.

Intuitively, a string of numbers will be called "random" if no pattern can be detected in it. Thus, *product randomness is lack of pattern.* Formalizing this intuition has proved challenging. Most of today's definitions are variants of the pioneering proposal of Per Martin-Löf, who in 1966 defined the randomness of a sequence in the language of algorithmic complexity. Without entering into the details, one ignores the actual process and defines randomness on a class of simulators, which are the algorithms that could have produced the sequence. In practice, product randomness is assessed through a battery of statistical tests. The product randomness of the outputs of a Bell test can of course be studied, but it doesn't have any special feature.

By contrast, a process is called "random" if one cannot predict how it will unfold. In other words, *process randomness is unpredictability.* Now we can make more precise the tight connection between process and product randomness: Unpredictability is not the same as lack of regularity, but a process that most of time produces a highly regular output will also be highly predictable. Conversely therefore, an unpredictable process will most likely produce a sequence that lacks regularity. Importantly, process randomness will almost always imply product randomness,[1] while the converse does not hold, as the example of the pre-recorded information shows. The reader can now guess which is *nonlocality's unique contribution: The possibility of certifying process randomness in a DI way, that is, without characterizing the process itself.*

The definition of process randomness calls for a predictor. The question "random for whom?" must always be addressed and is crucial to avoid confusions. Having noticed this, positing a proper formalization of process randomness is immediate: Randomness will be quantified by the *probability of the predictor guessing the output.*

8.1.2   Random for whom? The predictor

In a book devoted to nonlocality, defining the predictor may look like an easy task: Have we not proved that the outputs of some natural processes are random *for everyone*? A quick thought back to section 1.6 shows that "everyone" is rather ill-defined. For instance, any deterministic interpretation will be able to describe a hypothetical agent for whom there is no randomness. In the Bohmian interpretation, that would be an agent who has access to both the initial conditions and the quantum potential with infinite precision; in the many-worlds interpretation, an agent who perceives the whole multiverse and not just one of its branches. Also, there can't be randomness for a hypothetical agent who sees what for us is "the future" and has therefore nothing to predict.[2] So, a definition of the predictor is indeed needed. In this subsection and the next, we go through a list of characteristics, that will be summarized in Table 8.1.

---

[1] A formalized version of this statement is called Yao's theorem, see Theorem 11.9 in (Arora and Barak, 2009).

[2] This does not seem to prevent some people, who believe exactly that of God, from asking me quite often if quantum phenomena are random *for Him*.

Table 8.1 *The characteristics of the adversary Eve in this text. They capture a scenario of generation of secret randomness inside a secure location.*

Characteristics of Eve	Remarks
Is outside the secure location and has not tampered with anything inside it.	
Knows with certainty all that happens inside the secure location, apart from the outputs of the process under study (whose randomness has to be assessed) and possibly the content of a string of fixed finite length.	The fixed-length string is needed in the framework of randomness expansion and for QKD (section 8.4).
Has unbounded computational power and laboratory facilities	That is, she can run perfect simulations of the process, either digital or analog.
Has the most precise description of the process under consideration using quantum theory	In particular, she may have a pure-state description of each round of the process. Further relaxed in section 9.3.
Does not have quantum side-information.	Natural in the case of a single secure location. Most of the specialized literature gives quantum side-information to Eve (references in Appendix F.2.1).

First, we assume that *the predictor knows a perfect description of the process* and may run perfect analog or digital simulations with unbounded computational power. Such a predictor is not limited by what we describe as ignorance: If we describe the process with a mixed state, she may have a pure-state description of every round. She may also know exactly the misalignment of our devices, in which rounds the detectors in the lab will not fire . . .

But what is a "perfect" description, and which technologies can the predictor master? We assume that *the predictor bases her knowledge and technologies on quantum theory* (this will be relaxed in section 9.3).

An even more powerful predictor model has been considered in the literature: One that may hold the degree of freedom that purifies the state used in the process (the "purification"). In the jargon, we say that such a predictor has access to *quantum side-information*, while our predictor has only classical (though perfect) side-information. To appreciate why quantum side-information may help the predictor, we need to recall that randomness is not generated for the sake of it, but as a resource in further tasks. The predictor may not be interested the whole string of numbers: She may want to predict only those numbers that are used in a specific task, which may take place after some parts of the original string have been used or revealed. With classical side-information, the predictor can only update her knowledge based on Bayesian reasoning; holding on quantum side-information, she may be able to make a better guess.

8.1.3   The notion of secret randomness

Having defined process randomness, with the correlated notion of a predictor, the next distinction comes from the *use* to which the generated string is put. For some tasks, e.g., sampling and Monte Carlo optimization, the only property of randomness that is required is the lack of regularity. For such tasks, certifying product randomness would be sufficient, although certifying also process randomness cannot harm. By contrast, in cryptographic tasks the predictor is an *adversary* (traditionally called Eve). Process randomness is then necessary, because it guarantees that the generated string is still random for anyone who has not seen it. One then speaks of *secret randomness*.

Now, the possibility of secret randomness implies that the adversary should not be able to see the generated string. This requires the process to happen in a *secure location* to which the adversary has no access, be it by physical presence, or by hacking, or by the leakage of information, or by Trojan horses (Figure 8.1).

The necessity of postulating a secure location calls for two important remarks. First, one may ask: With Eve believed to be firmly and forever outside the single secure location, isn't just anything that happens inside "random for her"? This doubt stems from the most typical fallacy in dealing with randomness: Taking for granted that "things happen", i.e., that plenty of events are unpredictable for Eve. Postulating the existence of random processes inside the secure location would defeat the purpose of certifying another such a process. At times one postulates the existence of an *initial secret*, a string of *fixed* length that Eve ignores (of course, she knows that it exists and how it will be used). For the rest, apart from the outputs of the one process whose randomness we want to certify, all

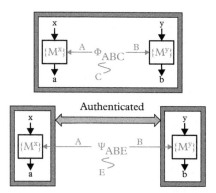

Figure 8.1 *Secret randomness requires secure locations. Top: The natural setting for randomness generation is that of a single secure location. The adversary Eve knows a perfect description of all the processes happening inside the location, but cannot tamper with them; the goal is to guarantee that (a, b) are random for Eve. Bottom: The setting for quantum key distribution involves two secure locations. In the space between the two, Eve can only listen to the communication on the authenticated channel but can tamper with everything else. In particular, since she can modify the state, the inputs x and y must also be random for her: In this setting, one speaks of randomness (or secret key) expansion.*

that takes place inside the secure location is supposed to be known to Eve, including the possible actions of human actors.

Second, the security of a location can never be absolutely guaranteed. This seems at odds with device-independent certification: Indeed, it has been the object of confusions in the literature (Appendix F.3). To avoid these confusions, we have to distinguish the certification of randomness from that of secret randomness. *Process randomness can be certified in a device-independent way, based on nonlocality*, and we are going to devote this chapter to it. But in order to use the generated string in cryptographic tasks (i.e., in order to claim that this randomness is *secret*) one needs to add the assumptions of secure location. Some of these assumptions bear on the devices: Notably, the devices must not have been fabricated by the adversary, otherwise she might have implemented a Trojan horse.[3] In this sense, it is fair to say that secret randomness cannot be certified in a fully DI way; but because the assumptions of secure location are required in any secrecy scenario, it is also fair to say that the DI certification of the process minimizes the total number of assumptions.

8.1.4   The disruptive role of nonlocality

Let us now go through different RNGs that could be in the secure location and discuss if they can be certified to be random for our Eve.

If the RNG is an algorithm, it should be rightly called *pseudo-RNG*: Obviously it has no process randomness for our Eve. The only possible randomness is an initial secret, which could be used as input of the algorithm. Nonetheless, however long the output string, the randomness remains only that of the initial secret.

If the RNG is a physical process that can be allegedly described with a *deterministic model*, there is still some hope for randomness against our Eve. On the one hand, the deterministic model may not be exactly accurate; on the other hand, even if it is accurate, it may require an enormous computational power. In both cases, an estimate of process randomness can only be based on trust of the modeling: In the first case, to describe the non-deterministic correction; in the second case, to be convinced of its irreducible complexity (even if one is willing to limit Eve's computational power, which we won't do here, one should be sure that Eve has not found a simpler description).

RNGs based on physical processes that require a quantum description are called *quantum random number generators (QRNGs)*. Here, even in very simple systems there is randomness for an adversary using quantum theory. Suppose that the accurate description of the process is a qubit in the state $|+\hat{z}\rangle$ being measured in the eigenbasis of $\sigma_{\hat{x}}$: Eve, with all her knowing this, can only despair. However, if the QRNG produces a local behavior as in this example, the device could contain a classical simulation (and Eve would know it). Once again, the estimate of process randomness requires

---

[3] To paraphrase a pun by Charles Bennett: Even with DI certification, secrecy devices should be DIY (do-it-yourself). The level of paranoia is a free parameter: For instance, is it enough to assume that the adversary was not involved in the assembling of the devices? Or should one make sure that even the screws have been fabricated inside the secure location?

trust in the modeling.[4] It's only for QRNGs based on nonlocality that a lower bound on process randomness can be obtained without any need for modeling the process.

Importantly, a Bell test requires inputs. These inputs must certainly be "random for the source" because we are assessing nonlocality assuming measurement independence (this will be partially relaxed in chapter 11, but some randomness will have to be kept). Even though measurement independence cannot be proved in absolute terms, there are many reasonable choices for uncorrelated processes, as discussed in subsection 1.5.3.

For the case of secret randomness, a subtler issue is whether the inputs of the Bell test must be random for Eve:

- If Eve can prepare the state, tamper with it, or hold quantum side-information, the inputs must be random for her. One then needs an initial secret in the secure location, and the process is useful only if more randomness is produced per round than is consumed to choose the inputs. We speak of *randomness expansion*.

- As described in Figure 8.1, the natural setting for randomness is that of a single secure location. In this setting, Eve cannot tamper with the state, and it is rather natural to assume that she does not hold a purification either. It is then possible to use inputs that are known to her, and speak of *randomness generation*.

Most of the literature deals with randomness expansion, because cryptographers find it always safe to give more power to the adversary. But randomness generation is simpler to address and its adversarial model is very reasonable: So in this book we stick to it, leaving a quick review of the other results for Appendix F.2.1.

8.2 Quantification of Randomness

8.2.1 Guessing probability: Definition

We formalize the definition of process randomness for i.i.d. processes, leaving for subsection 8.2.3 the discussion on how to remove this assumption.

In this general introduction to randomness, let us call Claire the player instead of our usual Alice and Bob. Let $\mathcal{C}$ be the set of outputs c of the process; its cardinality is denoted $|\mathcal{C}|$. Over several rounds, Claire observes the probability distribution $P(c)$. The adversary Eve however knows a more precise description ξ of the process in each round, so her knowledge of the output is described by $P_\xi(c)$. The best guess for Eve is obviously the most probable output given her knowledge:

$$c(\xi) = \text{argmax}_c P_\xi(c) \text{ [i.e., } P(c(\xi)) \equiv \max_c P_\xi(c) \text{]}. \tag{8.1}$$

---

[4] This does not mean that there is no advantage in choosing QRNGs: For instance, getting convinced that one is measuring the polarization of a photon is probably simpler than certifying that a process is irreducibly complex. But ultimately, which RNGs are better will depend on practical criteria: Simplicity, stability, cost, speed …

Let us denote by $q(\xi)$ the frequency at which the process ξ occurs: The joint probability distribution between Claire's and Eve's symbols is

$$P_{CE}(c, e) = \sum_\xi q(\xi) P_\xi(c) \delta_{e, c(\xi)} \tag{8.2}$$

Eve's average *guessing probability* will then be given by

$$P_{\text{guess}} = \sum_{c \in \mathcal{C}} P_{CE}(c, c) = \sum_\xi q(\xi) \max_c P_\xi(c). \tag{8.3}$$

Its maximal value is $P_{\text{guess}} = 1$ obtained if $P_\xi(c) = \delta_{c, c(\xi)}$ for all ξ. Its minimal value is $P_{\text{guess}} = \max_c P(c)$, obtained if the knowledge of ξ does not help Eve's guessing. The guessing probability for a N-symbol string is denoted $P_{\text{guess}}(N)$. It is equal to $(P_{\text{guess}})^N$ for a i.i.d. process.[5]

The guessing probability is the object that we shall set out to compute, or more precisely upper bound, based on nonlocality. Next, let us present its operational meaning.

8.2.2   Guessing probability and randomness extraction

The direct output of an alleged random process is termed the *raw output string*. Even if it contains some randomness, i.e., $P_{\text{guess}} < 1$, Eve may have a lot of information about it. Besides, Claire does not know what that information is: Eve may have guessed very well some rounds (for instance if there exist some ξ for which c is deterministic); or perhaps Eve's information is the same in each round ... At any rate, Claire would be ill advised to use that string for cryptographic purposes. She should rather *extract* the existing randomness into a shorter string such that her symbols are uniformly distributed and Eve is uncorrelated, that is $P'_{CE}(c', e') \approx \frac{1}{|\mathcal{C}|} P'_E(e')$.

The study of extraction constitutes a large body of theoretical and practical knowledge that goes far beyond the scope of a book on nonlocality.[6] The only role of nonlocality is to provide an estimate of the guessing probability. So we focus on the operational meaning of the guessing probability for extraction, leaving aside all the matters related to the design of extractors. To this effect, we notice that any extraction procedure can at best preserve the amount of randomness, so the length ℓ of the extracted string must be bounded by the requirement that $P'_{\text{guess}}(\ell) \geq P_{\text{guess}}(N)$. By definition, after the extraction, one should have $P'_{\text{guess}}(\ell) = |\mathcal{C}|^{-\ell}$. Thus

---

[5] Technical note: Not to be confused with a *mixture of i.i.d. processes*, an object that appears in De Finetti-type theorems. In that case, $P_{\text{guess}}(N) = \sum_\xi q(\xi)(P_{\xi, \text{guess}})^N$.

[6] The reader in need of a first introduction may appreciate the first section of (Pivoluska and Plesch, 2014). We also highlight that only some special extractors can be used if Eve has quantum side-information (De *et al.*, 2012).

$$\ell \leq -\log_{|\mathcal{C}|} P_{\text{guess}}(N) \equiv H_{\min}(C|E) \qquad (8.4)$$

which constitutes a definition of the *conditional min-entropy*.

Now, there exist extraction procedures that achieve equality in (8.4) in the asymptotic limit $N \to \infty$. The basic result goes by the improbable name of *leftover-hash lemma*, and there have been improvements. We have thus the operational interpretation to the min-entropy, or the guessing probability, as quantifier of randomness: It is a parameter that enters the construction of the extractor, specifying the optimal length of the extracted string.

Importantly though, all these procedures are based on *seeded extractors*. This means that the extraction procedure must be chosen among a set of possible ones using a string uncorrelated from the process and also random for Eve. If this independent randomness is not available, extraction may be reduced or even impossible: We'll discuss one example in section 11.5. The canonical example of deterministic (i.e., seedless) extraction is *von Neumann extraction*. It works only in the simplest case: The process must be a i.i.d. biased coin, $P(0) = p$ and $P(1) = 1 - p$ in each round, with $\frac{1}{2} < p < 1$. Besides, if one wants to extract secret randomness, von Neumann extraction works only if Eve does not have a better description of the process. In other words, we start from $P_{CE}(c, e) = P(c)\delta_{e,0}$ and $P_{\text{guess}} = p$ in all rounds. The two-symbol sequences $P(c_1 c_2)$ have probability $P(00) = p^2$, $P(01) = P(10) = p(1 - p)$ and $P(11) = (1-p)^2$. Claire discards the sequences 00 and 11, replaces the sequence 01 by 0 and the sequence 10 by 1. The length ℓ of the final string depends on the initial string, on average $\langle \ell \rangle = Np(1 - p)$. Notice that this is smaller than the min-entropy, which is $N(-\log_2 p)$ for $p > 1 - p$. Nonetheless, extraction has been achieved: For any given ℓ, the final distribution is $P'(c', e') = \frac{1}{2^\ell} P(e')$ for all c', whatever Eve does to her data to obtain e'.

8.2.3   Randomness in i.i.d. strategies and beyond

If Claire's data are compatible with the i.i.d. assumption, intuition suggests that a description in terms of a i.i.d. strategy should be fine.[7] For probability distributions, this intuition was put on rigorous mathematical grounds in the famous *De Finetti theorem*. This theorem, that can be applied to conditional processes too, was later generalized to states in quantum theory. For these developments I refer the reader to the masterful review by Renner (2007).

However, it took more effort before the suitable tool for DI certification was found.[8] The breakthrough came with the *entropy accumulation theorem* (Arnon-Friedman et al., 2019), which controls the convergence of *entropic quantities* towards values that would

---

[7] This is particularly clear in a Bayesian approach: If Claire starts with a i.i.d. prior belief, and the updates do not challenge it, then there is no reason to change that belief.

[8] The quantum De Finetti-type theorems depend on the knowledge of the Hilbert space dimension: In DI, one does not want to bound the dimension, thus the asymptotic convergence cannot be guaranteed. A De Finetti-type theorem for behaviors was eventually found, but it brings into play *signaling* i.i.d. behaviors, so once again it is not suitable for DI certification (Arnon-Friedman and Renner, 2015).

be obtained with i.i.d. strategies. This is sufficient for randomness and quantum key distribution, that rely indeed on entropies, though it cannot be used for self-testing for instance. Therefore, since here, we present only asymptotic results, for the remainder of this chapter we can work under the i.i.d. assumption, knowing that those results will hold in general.

8.2.4 Warm up: Randomness in characterized quantum systems

The quantification of process randomness is not among the standard calculations of quantum theory. This is why, before going to the device-independent estimates that are proper for nonlocality, it useful to study process randomness for *characterized* devices. Claire starts by characterizing the single-round state ρ through tomography (deviations from an exact i.i.d. description of the state can be taken care of by De Finetti-type theorems, since the dimension is known). Knowing the state, she performs suitable measurements. We will look at two cases.

The first case is that of Claire using the same measurement $\mathcal{M} = \{\Pi_c, c \in \mathcal{C}\}$ in all rounds. Her statistics are obviously given by $P(c) = \mathrm{Tr}(\rho \Pi_c)$. Eve's knowledge in any given round is described by a ρ_ξ, with $\rho = \sum_\xi q_\xi \rho_\xi$. An upper bound $\bar{P} \geq P_{\text{guess}}$ is obtained if the decomposition is the most favorable for Eve's guessing. By linearity of quantum theory, all the processes that lead to guessing the same output c can be grouped in a single sub-normalized state

$$\rho_c = \sum_{\xi \mid c(\xi) = c} q_\xi \rho_\xi. \tag{8.5}$$

Thus, the desired upper bound on P_{guess} will be the solution of

$$\begin{aligned} \bar{P} = \max_{\{\rho_c\}} \mathrm{Tr}\left(\sum_c \Pi_c \rho_c\right) \\ \text{subject to} \quad \rho_c \geq 0 \text{ for all } c \in \mathcal{C}, \\ \sum_c \rho_c = \rho. \end{aligned} \tag{8.6}$$

This is a semi-definite program (SDP) like those we encountered in section 6.3 and it can be solved by the same algorithms (recall in particular that the solution is guaranteed to be correct within numerical precision).

Here comes a case study. Claire knows she has a qubit and, has characterized its state as being $\rho = \frac{1}{2}(\mathbb{I} + \eta \sigma_{\hat{z}})$. She performs the projective measurement of $\sigma_{\hat{x}}$, that is $\Pi_\pm = |\pm \hat{x}\rangle\langle\pm \hat{x}|$. Her outputs are unbiased: $P(c = +1) = P(c = -1) = \frac{1}{2}$. But Eve can do better: The solution of (8.6) yields the upper bound $\bar{P} = \frac{1}{2}(1 + \sqrt{1 - \eta^2})$, obtained for $\rho_\pm = \frac{1}{2}|\chi_\pm\rangle\langle\chi_\pm|$ with $|\chi_\pm\rangle = \sqrt{(1 + \eta)/2}\,|+\hat{z}\rangle \pm \sqrt{(1 - \eta)/2}\,|-\hat{z}\rangle$ (more details in Appendix G.5, where we also show an example of the advantage of keeping quantum side-information).

As a second case study, we look at what happens if Claire alternates among two or more measurements. Our Eve knows which measurement is being performed in each round. However, measurement independence requires her description of the process to be uncorrelated from the measurement that is being performed; and *no* quantum description can determine all the outputs of a family of *incompatible measurements*. Thus, in addition to possible randomness in the state, there is an additional randomness coming from the measurements. A striking example is the measurement of a qubit in the maximally mixed state $\rho = \frac{1}{2}\mathbb{I}$ along two or three mutually unbiased bases: Using a simple generalization of (8.6), the best guessing probabilities are found to be $\bar{P} = \frac{1}{2}(1 + \frac{1}{\sqrt{2}})$ and $\bar{P} = \frac{1}{2}(1 + \frac{1}{\sqrt{3}})$ respectively[9] (Law *et al.*, 2014).

As a last remark: In these two examples, we have given both the state and the measurements. It may seem more natural to give the state and look for the optimal measurements. This optimization is no longer an SDP, so numerical methods can provide only heuristic bounds. In some simple cases, the exact solutions are known analytically. For instance, the choice of measurements in the two examples just mentioned are provably optimal; whereas the optimal solution is not known if Claire can alternate among four or more measurements on a qubit.

8.3 Device-Independent Certification of Randomness

We are now in a position to tackle the central section of this chapter: The DI certification of the amount of randomness. In Appendix F.2.1 the reader can find more information about the origin and development of the field, as well as further references.

8.3.1 Generic formulation

In a Bell scenario, Claire is replaced by Alice and Bob (and others if the scenario is multipartite, which we won't consider here). There should be some randomness for every pair of inputs and it may be advantageous to extract it all (Bancal *et al.*, 2014); but for simplicity, here we discuss only the randomness of the outputs $c \equiv (a_0, b_0)$ for the pair of inputs $x = y = 0$.

Alice and Bob see the behavior $\mathcal{P}$. Eve knows the actual behavior in each round, one of the $\mathcal{P}_\xi$ with $\mathcal{P} = \sum_\xi q_\xi \mathcal{P}_\xi$. Since convex combinations of behaviors are still behaviors, in analogy with what we did in subsection 8.2.4, we can group together all the processes that lead to guessing the same output in a sub-normalized behavior:

$$\mathcal{P}_{(a,b)} = \sum_{\xi|c(\xi)=(a,b)} q_\xi \mathcal{P}_\xi. \tag{8.7}$$

---

[9] By rewriting the result for two bases in terms of min-entropies, one recovers a state-independent entropic uncertainty relation derived in the pioneering work of Maassen and Uffink (1988).

Besides, by definition $\max_{(a',b')} P_{(a,b)}(a',b'|0,0) = P_{(a,b)}(a,b|0,0)$. Thus, the desired upper bound on P_{guess} will be the solution of

$$\begin{aligned} \bar{P} = \max_{\{\mathcal{P}_{(a,b)}\}} \sum_{(a,b)} P_{(a,b)}(a,b|0,0) \\ \text{subject to} \quad \sum_{(a,b)} \mathcal{P}_{(a,b)} = \mathcal{P} \\ \mathcal{P}_{(a,b)} \in \mathcal{Q} \text{ for all } a \in \mathcal{A}, \ b \in \mathcal{B} \end{aligned} \qquad (8.8)$$

The last line forces Eve's descriptions to be quantum behaviors, possibly sub-normalized; we have absorbed into it the positivity condition $P_{(a,b)}(a,b|x,y) \geq 0$ for all a, b, x, y. Without a constraint on the acceptable set of behaviors, the trivial upper bound $\bar{P} = 1$ is always achievable (if $\mathcal{P}$ is nonlocal, some $\mathcal{P}_{(a,b)}$ will have to be signaling). As we have seen in section 6.3, the membership in the quantum set is not easy to handle but it can be approximated by semi-definite criteria. Besides, if $\mathcal{Q}$ is replaced by a $\mathcal{Q}_{\mathcal{F}} \supset \mathcal{Q}$ as defined there, the optimization will lead to $\bar{P}_{\mathcal{F}} \geq \bar{P}$, so the bound remains a valid upper bound.

8.3.2 Case study: Randomness in Alice's output from CHSH

As a case study, we consider a CHSH test that has led to an observed value $S = S_{\text{obs}} > 2$. If $S_{\text{obs}} > 2$, there is randomness both in Alice's and in Bob's outputs. We are going to find an upper bound for the probability of guessing *one of Alice's outputs*, say a_0. The optimization to be performed is

$$\begin{aligned} \bar{P} = \max_{\mathcal{P}_+,\mathcal{P}_-} P_+(a=+1|x=0) + P_-(a=-1|x=0) \\ \text{subject to} \quad S[\mathcal{P}_+ + \mathcal{P}_-] = S_{\text{obs}}, \\ \mathcal{P}_+, \mathcal{P}_- \in \mathcal{Q}. \end{aligned} \qquad (8.9)$$

Notice that the only constraint related to observation is the value of S: We are not requesting $\mathcal{P}_+ + \mathcal{P}_-$ to be equal to the observed behavior.

This optimization can be cast as an optimization over a single behavior:

$$\begin{aligned} \bar{P} = \max_{\mathcal{P}'} P'(a=+1|x=0) \\ \text{subject to} \quad S[\mathcal{P}'] = S_{\text{obs}}, \\ \mathcal{P}' \in \mathcal{Q} \end{aligned} \qquad (8.10)$$

Indeed, define $P'_-(a,b|x,y) = P_-(-a,-b|x,y)$. Clearly, $\mathcal{P}'_- \in \mathcal{Q}$ if $\mathcal{P}_- \in \mathcal{Q}$; besides, since flipping both outputs does not change the correlations and CHSH depends only on correlations, we have $S[\mathcal{P}'_-] = S[\mathcal{P}_-]$. The form (8.10) is then obtained by letting $\mathcal{P}' = \mathcal{P}_+ + \mathcal{P}'_-$. We stress that the single-behavior optimization was not written *a priori*: It was derived from the rigorous optimization (8.9) using the symmetry of CHSH (Exercise 8.1).

Now we proceed to solve (8.10) analytically, following (Pironio *et al.*, 2010). Recall that we can work with projective measurements and pure states without loss of generality, since the dimension of the Hilbert spaces are not fixed and the purification has not leaked out from the secure location. Now we can use *Jordan's lemma*, which we have already mentioned in the previous chapter and is proved in Appendix G.4: If A_0 and A_1 are Hermitian operators on an arbitrary Hilbert space $\mathcal{H}$ with eigenvalues -1 and $+1$, there exist a basis in which both operators are block-diagonal

$$A_x = \bigoplus_\alpha A_x^{(\alpha)} = \bigoplus_\alpha \hat{a}_{x,\alpha} \cdot \vec{\sigma}^{(\alpha)} \tag{8.11}$$

with blocks of size at most 2×2 (the subspaces in which both operators commute will be characterized by $\hat{a}_{0,\alpha} = \pm\hat{a}_{1,\alpha}$). Of course, the same holds for Bob:

$$B_y = \bigoplus_\beta B_y^{(\beta)} = \bigoplus_\beta \hat{b}_{y,\beta} \cdot \vec{\sigma}^{(\beta)}. \tag{8.12}$$

The unknown state can then be written $|\Psi\rangle = \sum_{\alpha,\beta} \sqrt{p_{\alpha\beta}} |\psi^{\alpha\beta}\rangle$ where $|\psi^{\alpha\beta}\rangle$ is a normalized two-qubit state and $\sum_{\alpha,\beta} p_{\alpha\beta} = 1$. Also,

$$S_{obs} = \sum_{\alpha,\beta} p_{\alpha\beta} \langle \psi^{\alpha\beta} | \mathbf{S}^{\alpha\beta} | \psi^{\alpha\beta} \rangle. \tag{8.13}$$

with $\mathbf{S}^{\alpha\beta} = \sum_{xy}(-1)^{xy} A_x^{(\alpha)} \otimes B_y^{(\beta)}$. Finally,

$$P'(a = 0 | x = 0) = \sum_{\alpha,\beta} p_{\alpha\beta} \langle \psi^{\alpha\beta} | \Pi_0^{0\alpha} \otimes \mathbb{I}^\beta | \psi^{\alpha\beta} \rangle. \tag{8.14}$$

The optimization (8.10) is now cast as the maximization of a convex sum $P = \sum_k p_k P_k$ under a convex constraint $S = \sum_k p_k S_k \equiv S_{obs}$. Let's suppose that we find the solution for the elementary problem $\max_{S_k = s} P_k = f(s)$. If in addition $f''(s)$ does not change sign, we have two possibilities:

(i) If f is convex, the solution to the main problem is simply $P = f(S_{obs})$ obtained by setting all the $S_k = S_{obs}$.

(ii) If f is concave, the solution to the main problem is given by the convex combination of boundary points. For our problem, the boundary $S = 2\sqrt{2}$ has $P = \frac{1}{2}$, the boundary $S = 2$ has $P = 1$; so, if f were concave, with $S_{obs} = p2\sqrt{2} + (1-p)2$ the solution would be $\bar{P} = p\frac{1}{2} + (1-p) = 1 - \frac{1}{2}(S_{obs} - 2)/(2\sqrt{2} - 2)$.

So, our next step is to find the maximal value of $\langle \psi | \Pi_0^0 \otimes \mathbb{I} | \psi \rangle$ over the set of *pure two-qubit states* for a given value $S = s$. From (3.17), we know that the pure state $|\Psi(\theta)\rangle = \cos\theta |00\rangle + \sin\theta |11\rangle$ can reach $s = 2\sqrt{1 + \sin^2 2\theta}$. By writing down the projector for that state, we see that its most biased marginal is $P(+|\hat{z}) = \frac{1}{2}(1 + \cos 2\theta)$. States with smaller values of θ can also reach the same s (for suboptimal measurements), but their most biased marginal is definitely lower; states with higher values of θ cannot reach s.

Therefore, by solving explicitly for θ, we find, after trivial algebra, $f(s) = \frac{1}{2}(1 + \sqrt{2 - (s/2)^2})$. This function being convex, we have found the solution to the general problem:

$$\bar{P}(S_{\text{obs}}) = \frac{1}{2}\left[1 + \sqrt{2 - \left(\frac{S_{\text{obs}}}{2}\right)^2}\right]. \tag{8.15}$$

Two remarks can be made at this stage:

- There is an interesting connection with self-testing. Let us take the behavior that gives the highest violation of CHSH for a non-maximally entangled two-qubit state $|\Psi(\theta)\rangle$, having chosen the measurements such that $A_0 = \sigma_{\hat{z}}$. In this case we would have $P(a = 0|x = 0) = \bar{P}(S_{\text{obs}})$. But the r.h.s. is an upper bound on Eve's guessing probability, while the l.h.s. is part of the behavior observed by Alice. Thus, Eve's knowledge does not allow her to guess better than Alice. For the power given to Eve, we expect this to be the case only when she knows that the state is pure. In fact, one can prove that the observation of both $S[\mathcal{P}] = S_{\text{obs}}$ and $P(a = 0|x = 0) = \bar{P}(S_{\text{obs}})$ self-tests $|\Psi(\theta)\rangle$ (Exercise 8.2).
- Having the guessing probability, one can compute the min-entropy (8.4), which is a strict lower bound on the amount of randomness. But in practice, a perfect parameter estimate (e.g., of S_{obs}) cannot be obtained from finite data; and the absolute worst-case assessment is always that there is no randomness, since any finite string may be the outcome of a local process, however unlikely. This is why one rather opts for deriving rigorous bounds for the *typical* min-entropy (Pironio and Massar, 2013; Arnon-Friedman *et al.*, 2018). For instance, for the randomness of a_0 as a function of the violation of CHSH one would get in the asymptotic limit

$$r_{\text{typ}} = 1 - h\left(\frac{1}{2} + \frac{1}{2}\sqrt{\left(\frac{S_{\text{obs}}}{2}\right)^2 - 1}\right) \tag{8.16}$$

where $h(p) = -p\log_2 p - (1 - p)\log_2(1 - p)$ is the binary Shannon entropy. Notice that $r_{\text{typ}} > -\log_2 \bar{P}(S_{\text{obs}})$ computed from (8.15): Even in the asymptotic limit, a typicality result is more optimistic than a strict one. We won't discuss these matters further here: They are very important for information processing and actual experiments [see e.g., (Shen *et al.*, 2018)], but they don't shed further light on the role of nonlocality.

8.3.3 Optimizing the amount of randomness

Among the many other results that have been obtained in DI randomness generation, we focus on the discussions about extracting the maximal amount of randomness. The pioneering paper is (Acín *et al.*, 2012), but we describe later results that encompass and clarify those ones.

We have just seen that the maximal violation of CHSH guarantees that one of Alice's outputs is maximally random: It's a bit with $P(0) = P(1) = \frac{1}{2}$. The first natural extension is to *extract randomness from one output of Alice and one of Bob*. These are two bits: For maximal randomness, $P(a_{\bar{x}}, b_{\bar{y}}) = \frac{1}{4}$ must hold for two suitable inputs $\bar{x}, \bar{y}$. In the behavior (3.11) that maximizes CHSH, however, all the probabilities are $\frac{1}{4}(1 \pm \frac{1}{\sqrt{2}})$. In order to extract two bits of randomness, one can add a third input $y = 2$ to Bob with the property $\langle A_0 B_2 \rangle = 0$ and $\langle A_1 B_2 \rangle = 1$. If $S = 2\sqrt{2}$ with $x, y \in \{0, 1\}$, we know that (up to local isometries) the state is $|\Phi^+\rangle$, A_0 is $\sigma_{\hat{z}}$ and A_1 is $\sigma_{\hat{x}}$. Given this, $\langle A_1 B_2 \rangle = 1$ implies that B_2 is $\sigma_{\hat{x}}$ on Bob; therefore $A_0 B_2$ acts as $\sigma_{\hat{z}} \otimes \sigma_{\hat{x}}$ on $|\Phi^+\rangle$, whence $P(a_0, b_2) = \frac{1}{4}$ as desired (Mironowicz and Pawłowski, 2013; Law *et al.*, 2014).

Next, having a maximally entangled state and/or the maximal violation of a tight Bell inequality is not the only way to get maximal randomness. Specifically, one can devise a Bell test that extracts two bits of randomness for *any non-maximally entangled state of two qubits*. From the state behavior (3.29), it is clear that this can be achieved only if $A_{\bar{x}}$ and $B_{\bar{y}}$ self-test two orthogonal directions in the $(\hat{x}, \hat{y})$ plane of the Bloch sphere. Having noticed this, the solution follows the same pattern as before: One self-tests $|\Psi(\theta)\rangle$ as well as the two suitable additional measurements.

But self-testing inspires further extensions. For the sake of the example, let us go back again to the self-testing of $|\Phi^+\rangle$: A qubit is being measured on Alice's side, and an extremal POVM on qubits has four outcomes. If one could self-test that an additional measurement $A_{\bar{x}}$ is a suitable such POVM, we would have $P(a_{\bar{x}}) = \frac{1}{4}$, i.e., we could extract two bits of randomness from Alice's certified qubit alone. Two explicit solutions have been given in (Acín *et al.*, 2016).

Shortly later it was proved that this is not the limit yet: One can in principle extract an *unbounded amount* of randomness from Alice's certified qubit in a device-independent way (Curchod *et al.*, 2017). The protocol builds on a remarkable observation (Silva *et al.*, 2015): If Alice's operation consists of a *sequence of suitable weak measurements* labeled by j, each of the behaviors $P(a_j, b|x_j, y)$ can violate CHSH. Such a sequence of measurements cannot be described as a single POVM if the choice of a later measurement may depend on the output of the previous ones; thus, this protocol is not bound to yield at most 2 bits of randomness.[10]

8.4 Device-Independent Quantum Key Distribution

8.4.1 Introduction to QKD

As mentioned at the beginning of Part II and in detail in Appendix F, it is by studying *quantum key distribution* (QKD) that the idea of device-independent certification came

---

[10] For characterized devices, the possibility of extracting unbounded randomness from a qubit is trivial even with a fixed sequence of strong measurements: One can first measure $\sigma_{\hat{z}}$, then $\sigma_{\hat{x}}$, then again $\sigma_{\hat{z}}$, then $\sigma_{\hat{x}}$ ... After the first measurement, the information about the initial state has been erased, and thus not more than one bit of randomness is extracted from the original state, compatible with the fact that the measurements are projective. The subsequent randomness comes from the incompatibility of measurements themselves, not from the initial state.

to the fore. It is fair to finish this part with some comments about this task. For readers who want to know more: The basic notions of QKD can be gathered from the review (Scarani *et al.*, 2009); to study DI security of QKD, it is good to start with the first such proof, that assumed an i.i.d. process (Acín *et al.*, 2007; Pironio *et al.*, 2009), then go straight to the application of the entropy accumulation theorem that vindicated the same asymptotic bound without that assumption (Arnon-Friedman *et al.*, 2019, 2018).

The goal of QKD is the generation of a secret key by two distant players, Alice and Bob. A secret key is a string of numbers that must be identical for Alice and Bob and are random for Eve. In a nutshell, the procedure consists of three steps:

1. By exchanging quantum signals, Alice and Bob generate their own raw strings of bits (*raw keys*).

2. By classical communication, they run an *error correction* protocol, at the end of which Alice's and Bob's strings are identical.

3. To remove Eve's information, they apply the same randomness extraction protocol on their strings. This step is called *privacy amplification*.

Thus, like in the case of randomness, the role of quantum theory (and of nonlocality in the DI framework) is to provide an estimate of min-entropy.

The task of QKD thus requires the existence *two separated secure locations, connected by two channels* (Figure 8.1, bottom). On the channel used to send quantum signals (*quantum channel*) Eve can tamper as much as she wants. The *classical channel* should be "authenticated": This means that Alice and Bob know that they are talking to one another, and Eve can only listen but not tamper (if Eve could impersonate Bob, Alice would end up establishing the secret key with her and the task would be impossible).

8.4.2   Differences between QKD and randomness

The setting of two separated secure locations, connected by the two channels just described, makes QKD more demanding than randomness generation. We finish this chapter by going through the main differences.

First, for the authentication of the classical channel, Alice and Bob need an initial shared secret. This is obvious without any knowledge of the technicalities of authentication: If I want to make sure that the person I am talking to is the right one, I must be able to request some identity token, previously agreed upon. Thus, *there is no key generation: Only key expansion.*

Second, on the quantum channel, Eve can tamper with the state. If she knew in advance which measurements are being performed in each round, she could realize a man-in-the-middle attack (a.k.a. intercept-resend): She would perform the measurements herself, learn the output and resend a signal to Alice and/or Bob prepared in the corresponding eigenstate. To avoid this, in QKD *the inputs of the process must be random for Eve.* Therefore, Alice and Bob must have some at least some initial randomness in their locations. Usually, it is assumed that there exist local random processes inside each location: For QKD, this does not defeat the purpose, since the desired product is not local randomness but a joint secret key.

For the same reason that she has access to the system, a technologically unbounded Eve can naturally hold *quantum side-information*. Bounded-storage models have been studied, that restrict Eve's power of holding to the purification, but usually they are not invoked in actual security proofs.

The most important difference is the issue of spatial distance. In randomness generation, the distance between Alice and Bob can be kept at the minimum (e.g., dictated by the locality loophole), which is just as well, since it all must fit in the same secure location. Conversely, in the case of QKD, the larger the distance between Alice and Bob, the more relevant the task. However, longer distances imply greater losses on the quantum channel; and if the losses are too high, the detection loophole cannot be closed. Even assuming that the detection loophole is closed, losses introduce errors in the raw strings of Alice and Bob, and error correction will consume part of the min-entropy. In other words, a loophole-free Bell test can be almost immediately read as DI randomness generation, but not as DI certification of QKD. For the latter, a direct link solution is probably not feasible and is certainly limited (see Exercise 8.3 for a rough estimate). Other solutions are being considered: For the state-of-the-art at the moment of writing, we refer to (Máttar *et al.*, 2018).

· ·

EXERCISES

Exercise 8.1 *Generalizing subsection 8.3.2, we want now to bound the randomness associated to the pair (a_0, b_0) of outputs of CHSH, knowing only the observed violation S_{obs}.*

(a) *Write down the optimization, analog of (8.9).*

(b) *Prove that the symmetries of CHSH allow the reduction to a two-behavior optimization, but not to a single-behavior optimization like (8.10).*

Remark: For CHSH, it was checked that the result of the two-behavior optimization is the same as that one of a single-behavior optimization. This observation points to a property of the solution that may provide some insight; but cannot be used to justify the use of a single-behavior optimization a priori.

Exercise 8.2. *At the end of subsection 8.3.2, we made the remark that the joint observation of S_{obs} given by (3.18) for CHSH, together with $P(a = 0|x = 0) = \bar{P}(S_{obs})$ given in (8.15), is a self-testing of the state $|\Psi(\theta)\rangle = \cos\theta\,|00\rangle + \sin\theta\,|11\rangle$. Prove that this is the case:*

(a) *By writing an ab initio proof along the same lines as the one in subsection 7.1.1.*

(b) *By proving that those observations match the self-testing criterion for that state given in (Bamps and Pironio, 2015): The maximal violation of the tilted CHSH inequality*

$$\alpha\,\langle A_0\rangle + \langle A_0 B_0\rangle + \langle A_0 B_1\rangle + \langle A_1 B_0\rangle - \langle A_1 B_1\rangle = \sqrt{8 + 2\alpha^2} \qquad (8.17)$$

for a particular $\alpha = \alpha(\theta)$ that you can deduce here.

Exercise 8.3. *The fraction of extractable secret key in a QKD protocol is estimated by the difference between the Shannon entropies of Eve and that of Bob on Alice's string [for precise statements and justification, refer to the review (Scarani et al., 2009)].*

We consider a QKD protocol implemented with a perfect maximally entangled state $|\Phi^+\rangle$ of two photonic qubits. Alice and Bob extract the key by measuring $\sigma_{\hat{z}}$. Eve's information is estimated in a DI way by checking the violation of CHSH with the optimal measurements for $|\Phi^+\rangle$.

We assume that both players have detectors with perfect efficiency, and that the source is in Alice's lab, so that she detects every photon. The photons traveling to Bob arrive with probability η; in the rounds when he does not detect, he chooses a random value for his bit.

1. *Prove that the predicted value of CHSH is $S = 2\sqrt{2}\eta$.*
2. *Prove that the error fraction in the raw key is $\epsilon = P(a_{\hat{z}} \neq b_{\hat{z}}) = \frac{1}{2}(1 - \eta)$.*
3. *The fraction of extractable secret key is therefore $r = h(\bar{P}(S)) - h(\epsilon)$ where $\bar{P}(S)$ is given by (8.15) and h is the binary Shannon entropy $h(p) = -p\log_2 p - (1 - p)\log_2(1 - p)$. Compute numerically the value of η such that $r = 0$.*
4. *An optimistic estimate for optical fibers is $\eta = 10^{-d/50}$ where d is the distance measured in kilometers. Compute what would be the critical distance for a fiber implementation of this protocol.*

Part III

Foundational Insights from Nonlocality

Nonlocality has always had a strong "foundational" flavor, manifested mainly in discussions about interpretations, but also in more technical works. The latter are the object of this last part. Being a phenomenon independent of quantum theory, nonlocality can be described in other mathematical frameworks. One can then discuss which features of nonlocality are robust to a possible change of theory, and which are so dependent on quantum theory that could be taken as part of its definition.

9

Nonlocality in the No-Signaling Framework

Cosí tra questa immensitá s'annega il pensier mio: e il naufragar m'é dolce in questo mare.

Amidst this immensity my thought drowns: And to flounder in this sea is sweet to me.

G. Leopardi, *L'infinito*

In this chapter and the next, we extend the study of Bell nonlocality beyond quantum behaviors to consider the broader set of no-signaling behaviors. After introducing the formalism, we show that some features of quantum theory are actually shared by any theory predicting no-signaling behaviors.

9.1 The No-Signaling Polytope

No-signaling (NS) behaviors have been already defined and studied in section 2.2, to which we refer. The set of NS behaviors forms a polytope, naturally called *the no-signaling polytope*. Indeed, the convex sum of two NS behaviors is another NS behavior, so the set is convex. Besides, once the NS constraints (2.9) are enforced, the possible behaviors are only limited by the positivity conditions: Thus, the set of NS behaviors is delimited by finitely many hyperplanes.

Because of this, contrary to what happens with the local polytope, the equations of the facets are easily listed, while some work is required to determine the extremal points. The local deterministic behaviors remain extremal points also for the no-signaling polytope. In addition to those, the polytope has *nonlocal extremal behaviors*. Clearly, these extremal behaviors are *not deterministic*, since the only deterministic NS behaviors are the local deterministic ones (recall Exercise 2.2). The highest violation of an inequality $I \leq I_L$ by the extremal points of the no-signaling polytope is called the no-signaling limit I_{NS}—recall that this may not coincide with the algebraic limit, as shown by the counterexample of the GYNI inequality in subsection 5.1.3.

The simplest example of nonlocal extremal NS behavior is called *PR-behavior* from Popescu and Rohrlich, who found it in previous works (Rastall, 1985; Khalfin and Tsirelson, 1985) and used it in a very influential work (described in chapter 10). We shall study this behavior in the next section. As it turns out, there is no other extremal

Bell Nonlocality. Valerio Scarani. © Valerio Scarani 2019. Published in 2019 by Oxford University Press.
DOI: 10.1093/oso/9780198788416.001.0001

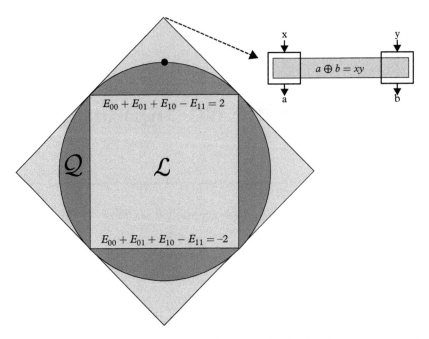

Figure 9.1 *The no-signaling polytope in the (2, 2; 2, 2) scenario. The drawing represents exactly the two-dimensional slice of the 8-dimensional object that contains four nonlocal extremal points (PR-behaviors). The curved line of the quantum set represents the TLM boundary (6.5), the black dot represents the behavior (3.11) that gives $S = 2\sqrt{2}$ for the form (2.29) of CHSH. On the side, a pictorial representation of the hypothetical PR-box in the same style as Figure 1.1.*

NS behavior above each CHSH facet; in other words, the no-signaling polytope of the CHSH scenario has only eight extremal nonlocal points, that are all equivalent to the PR-behavior up to the usual relabelings. The most frequent representation of this polytope is given in Figure 9.1; for plots of other slices with more intricate geometric features, see (Goh *et al.*, 2018).

Beyond the CHSH scenario, the extremal NS behaviors have been completely listed for the (2, *m*; 2, *m*) scenarios (Barrett *et al.*, 2005*b*) and the (*M*, 2; *M*, 2) scenarios (Barrett and Pironio, 2005; Jones and Masanes, 2005). In both cases, they are rather straightforward generalizations of the PR-behavior. Once again, this pleasant intuition fails when moving to multipartite scenarios. In the (2, 2; 2, 2; 2, 2) scenario (Pironio *et al.*, 2011), there are 45 inequivalent nonlocal extremal NS behaviors. We know from subsection 5.1.3 that there are also 45 inequivalent non-trivial facets of the local polytope. The equality in number is not accidental (Fritz, 2012), but does not lead to a simple geometry, on the contrary: Above any local facet one finds several inequivalent extremal NS points; and some of the extremal NS points do not achieve I_{NS} for any local facet.

9.2   The PR-Box

9.2.1   The PR-behavior and the hypothetical PR-box

We work in the CHSH Bell scenario. Rather than proceeding formally to find the intersection of the no-signaling facets, we introduce the PR-behavior in a heuristic way by asking what a behavior must satisfy in order to reach $S = 4$. The answer is clear: The correlation coefficients must satisfy $E_{00} = E_{01} = E_{10} = -E_{11} = 1$, that is[1]

$$a \oplus b = xy \quad \text{for} \quad a, b \in \{0, 1\} \tag{9.1}$$

where $a \oplus b = (a + b) \mod 2$. There are sixteen deterministic behaviors, obviously signaling, that satisfy (9.1):

$$P_\gamma(a, b|x, y) = \delta_{a=\gamma_{xy}} \delta_{b=\gamma_{xy} \oplus xy} \quad \text{with} \quad \gamma = (\gamma_{00}, \gamma_{01}, \gamma_{10}, \gamma_{11}) \in \{0, 1\}^4. \tag{9.2}$$

Let us prove that only one convex combination of these behaviors is no-signaling. Indeed, for $\mathcal{P} = \sum_\gamma q(\gamma) \mathcal{P}_\gamma$ the no-signaling conditions (2.9) read

$$P(a|0,0) = P(a|0,1) : \sum_\gamma q(\gamma)\delta_{a=\gamma_{00}} = \sum_\gamma q(\gamma)\delta_{a=\gamma_{01}},$$

$$P(a|1,0) = P(a|1,1) : \sum_\gamma q(\gamma)\delta_{a=\gamma_{10}} = \sum_\gamma q(\gamma)\delta_{a=\gamma_{11}},$$

$$P(b|0,0) = P(b|1,0) : \sum_\gamma q(\gamma)\delta_{b=\gamma_{00}} = \sum_\gamma q(\gamma)\delta_{b=\gamma_{10}},$$

$$P(b|0,1) = P(b|1,1) : \sum_\gamma q(\gamma)\delta_{b=\gamma_{01}} = \sum_\gamma q(\gamma)\delta_{b=\gamma_{11} \oplus 1}.$$

Now, take $a = b = \xi$: Enchaining the conditions implies that all the sums must be equal, and the requirement $\sum_\gamma q(\gamma)\delta_{\xi=\gamma_{11}} = \sum_\gamma q(\gamma)\delta_{\xi=\gamma_{11} \oplus 1}$ fixes their values to be $\frac{1}{2}$. Thus, in order to satisfy the no-signaling conditions, a convex combination of the $\mathcal{P}_\gamma$ must have unbiased marginals. This forces the unique solution[2]

$$\mathcal{P}_{PR} : P(a, b|0, 0) = P(a, b|0, 1) = P(a, b|1, 0) = \frac{1}{2}\delta_{a=0,b=0} + \frac{1}{2}\delta_{a=1,b=1},$$

$$P(a, b|1, 1) = \frac{1}{2}\delta_{a=0,b=1} + \frac{1}{2}\delta_{a=1,b=0}. \tag{9.3}$$

---

[1] The choice $a, b \in \{0, 1\}$ is almost universal in the literature, but of course is not constraining. If someone prefers to work with $a, b \in \{+1, -1\}$, the analog of (9.1) is $ab = (-1)^{xy}$.

[2] Indeed, for $xy = 0$, all $\mathcal{P}_\gamma$ are such that (a, b) is either $(0, 0)$ or $(1, 1)$: The requirement that Alice's (Bob's) output must be unbiased forces to take both possibilities with the same weight. Similarly for $xy = 1$, where (a, b) is either $(0, 1)$ or $(1, 0)$.

This uniqueness, together with the fact that $S = 4$ is the algebraic maximum of CHSH, implies that $\mathcal{P}_{PR}$ cannot be realized by mixing other no-signaling behaviors: It is an extremal point of the no-signaling polytope.

The PR-behavior *per se* is nothing more than the answer to a mathematical question. However, in the context of quantum information theory, a change of perspective took place, and several works focused on the properties of a *hypothetical resource* that would exhibit such a behavior. This resource came to be called *PR-box* from the way it was usually drawn (Figure 9.1). Several results quickly followed (Barrett *et al.*, 2005*b*; Barrett and Pironio, 2005; Jones and Masanes, 2005; Masanes *et al.*, 2006); it is beyond the scope of this book to review them in detail. Here we follow one story line: The conjecture that the PR-box could be the "unit of nonlocality"; the quest for the quantum principle that was the motivation of Popescu and Rohrlich (1994) is the subject of for chapter 10.

9.2.2 Feats and failures of the PR-box as unit of nonlocality

In entanglement theory (see Appendix C.2.3), the amount of entanglement is measured by *how many singlets* are needed to create a state (entanglement of formation), can be extracted from a given state (distillable entanglement), and so on. This is not a mere arbitrary choice. The resources are equivalent insofar as the transformation between singlets and any state is reversible in the limit of infinitely many copies. Also, with sufficiently many shared singlets, even multipartite states can be prepared with local operations and classical communication, through teleportation and entanglement swapping. The singlet is indeed the *unit resource* for entanglement.

Could the PR-box play a similar role of *unit of nonlocality*? It does share several analogies with the singlet. First and obviously, it is maximally nonlocal in the simplest scenario (2, 2; 2, 2), just as the singlet is maximally entangled for the simplest Hilbert space $\mathbb{C}^2 \otimes \mathbb{C}^2$. Then it is *monogamous* or, seen as dynamics rather than kinematics, a *no-cloning* theorem holds (Exercise 9.1). Possibly the most striking connection between the PR-box and the singlet was provided in (Cerf *et al.*, 2005). Consider the *singlet behavior*, i.e., the state-behavior that describes the statistics of all possible projective measurements on the singlet:[3]

$$\mathcal{P}_{\Psi^-} = \left\{ P(a, b|\hat{a}, \hat{b}) = \frac{1}{4}(1 - ab\,\hat{a} \cdot \hat{b}) \,\middle|\, a, b \in \{-1, +1\}, \hat{a}, \hat{b} \in \mathbb{S}^2 \right\}. \tag{9.4}$$

Cerf and coworkers proved that this behavior can be simulated exactly if in each round the players share a PR-box alongside with LVs. The explicit protocol is presented in Appendix G.6 following the approach of (Degorre *et al.*, 2005). This result is remarkable because the only nonlocal resource needed to reproduce the singlet behavior is a no-signaling process with binary input and binary outputs; the complexity of infinitely many

---

[3] It is of course the Werner behavior (3.24) for $W = 1$, and is related by local changes of basis to the state-behavior (3.29) for $\theta = \frac{\pi}{4}$.

settings on a sphere is taken care of by the local variables.[4] One is even tempted to go beyond simulations and conjecture that this may be what a singlet "really" is: A PR-box complemented with local variables. It was later proved that, in order to simulate weakly-entangled two-qubit pure states, the PR-box must be queried more than once in each round (Brunner *et al.*, 2005). This may not be an unwelcome feature for a unit of nonlocality, given that nonlocality and entanglement do not always behave monotonically (e.g., in the violation of some inequalities, section 4.3; the threshold efficiency for the detection loophole, Appendix B.3). That being said, no link was found between this result on simulations and those other observations.

Finally, for bipartite scenarios the PR-box lives up to the role of unit resource (Barrett and Pironio, 2005; Jones and Masanes, 2005): Given sufficiently many PR-boxes shared between two players, all the extremal NS behaviors of those scenarios (that is the whole no-signaling polytope) can be generated by local operations.

Unfortunately, the PR-box fails to play the same role when moving to multipartite Bell scenarios. Barrett and Pironio (2005) proved that some multipartite no-signaling behaviors can't be simulated if the players initially share only bipartite PR-boxes, even in arbitrarily large numbers. The core of their proof is instructive and can be sketched here. Consider the behavior in the $(2, 2; 2, 2; 2, 2)$ scenario defined by the five perfect correlations

$$a_0 \oplus b_1 = 0$$
$$b_0 \oplus c_1 = 0$$
$$c_0 \oplus a_1 = 0, \tag{9.5}$$
$$a_0 \oplus b_0 \oplus c_0 = 0$$
$$a_1 \oplus b_1 \oplus c_1 = 1$$

all the other expectation values being maximally mixed. This behavior is a nonlocal[5] extremal point of the no-signaling polytope. Because it's nonlocal, the players must rely on some nonlocal resource to simulate it. Our assumption is that they share only bipartite PR-boxes. Let's put ourselves in Alice's shoes, and suppose that in a given round she received the input $x = 0$. If Bob has received $y = 1$, Alice should get correlated only to him, and so she should not use the outputs of the PR-boxes she shares with Charlie. If Bob has received $y = 0$, Alice should be correlated with Bob and Charlie together: In this case, she cannot ignore the outputs of the PR-boxes she shares with Charlie. By symmetry, all players are in the same predicament: Thus, the behavior (9.5) can't be simulated with those shared resources. While this behavior can't be realized with quantum resources, a five-partite analog can, and the same kind of proof applies. So, with only shared bipartite PR-boxes one can't even reconstruct all the possible multipartite quantum behaviors.

---

[4] This work was inspired by the model of Toner and Bacon (2003), that achieves the same with the simplest *signaling* resource, namely one bit of communication (see section 11.2).

[5] By now, the reader should have noticed that the nonlocality is manifest: The first four correlatons would imply $a_1 \oplus b_1 \oplus c_1 = 0$ for LHV.

This observation suggests that it is impossible to define the analog of teleportation and entanglement swapping, because these are the protocols through which shared singlets can be used to generate any multipartite state. Indeed, the analog of a Bell state measurement for PR-boxes is problematic: It cannot be consistently defined over the whole no-signaling polytope (Skrzypczyk *et al.*, 2009). In general, PR-boxes have poor dynamics: The set of transformations that can be applied to these hypothetical resources without leaving the framework is very limited (Barrett, 2007; Dall'Arno *et al.*, 2017).

9.2.3 An "implausible consequence" of having a PR-box

PR-boxes may fall short of all the desiderata for a unit of nonlocality, but would nonetheless be very powerful resources. The first[6] to highlight their power was Wim van Dam, in the context of *communication complexity* (van Dam, 2013). Consider the following game: The verifier draws two strings of n bits $\underline{x}, \underline{y} \in \{0,1\}^n$ uniformly at random, sends $\underline{x}$ to Alice and $\underline{y}$ to Bob. Contrary to the rules of a Bell test, here the players can exchange m bits of communication. The goal is for Alice to output the value of a function $f(\underline{x},\underline{y})$.

Any such game can be won if $m = n$, because Bob can send $\underline{y}$ to Alice. For some choices of f, a much smaller amount of communication is sufficient: For instance the parity sum $f(\underline{x},\underline{y}) = \bigoplus_{k=1}^{n}(x_k \oplus y_k) = X \oplus Y$ with $X = \bigoplus_{k=1}^{n} x_k$ and $Y = \bigoplus_{k=1}^{n} y_k$. Bob can compute locally the parity Y of his own string and send this single bit to Alice, who can then compute f. But it is not surprising and can be rigorously proved that, given a suitable choice of f, the game can be won perfectly *only* if the players can exchange n bits. One such choice is the inner product $f(\underline{x},\underline{y}) = \underline{x} \cdot \underline{y} = \bigoplus_{k=1}^{n} x_k y_k$.

However, if the players share a PR-box, *one single bit* of communication would be sufficient to win the game perfectly. Indeed, if Alice and Bob input sequentially all the (x_k, y_k) into the PR-box, due to (9.1) the corresponding outputs (a_k, b_k) will satisfy $\bigoplus_{k=1}^{n} x_k y_k = \bigoplus_{k=1}^{n}(a_k \oplus b_k)$. In other words, the PR-box has transformed products of bits into sums. At this point, to win the game Bob needs to send the single bit $B = \bigoplus_{k=1}^{n} b_k$ to Alice.

Now, any Boolean function can be written as a sum of products. Thus, the PR-box *trivializes communication complexity*, which van Dam deemed "implausible consequence," all the more because shared entanglement would not lead to any improvement.[7] This work paved the way for a series of studies with a much more ambitious scope: Finding implausible consequences not just for the PR-box, but for every hypothetical no-signaling resource that produces behaviors outside the quantum set. This will be the topic of chapter 10.

---

[6] Van Dam noticed this point in his Ph.D. thesis, defended in Oxford in the year 2000, and posted it on the arXiv in 2005, but was discouraged to submit it for publication by more senior scientists. The argument is almost trivial indeed, but it became very influential, as we shall see in chapter 10. It was eventually published in 2013, as cited in the text. Nobody was harmed: The result was accessible, priority has always been recognized, and Wim holds a permanent position.

[7] Entanglement does help in some communication complexity tasks, even providing an exponential reduction of the required communication; but not for the inner product function (Buhrman *et al.*, 2010).

9.3 Device-Independent Certification Reloaded

Device-independent certification can now be revisited by relaxing the requirement that the adversary's knowledge and technical skills are based on quantum theory (subsection 8.1.2). Now, our adversary masters a theory of no-signaling resources, in which all no-signaling behaviors can be produced.

9.3.1 Randomness generation

For randomness generation, the adversary can describe the process with her theory of no-signaling resources, and can run a simulation based on such resources—for short, we have an *adversary with no-signaling description*.

Under the i.i.d. assumption, the study of randomness is simpler than in the quantum case. Since the NS set is a polytope, the constraint that behaviors must be no-signaling is a linear constraint, so the analog of the optimization (8.8) becomes a linear program. Even more, that constraint can be implemented automatically by decomposing the behavior on the set of extremal behaviors, which are finitely many and known—there is still an optimization to be performed, since the number of extremal behaviors is larger than D_{NS}, thus a generic behavior may admit several decompositions. The result of the optimization is a *tight* upper bound[8] on P_{guess} for our Eve.

In the (2, 2; 2, 2) scenario, the study of randomness against an adversary with no-signaling description is almost trivial due to the geometry of the NS set. Recall that there are only eight extremal nonlocal behaviors, all versions of the PR-behavior (9.3). A decomposition of the observed behavior into extremal one determines the fraction q_{PR} of rounds in which the nonlocal behaviors are used. In those rounds, there is randomness for Eve, whereas she knows everything when deterministic behaviors are used. The optimization thus consists in minimising q_{PR}. Given $S_{obs} > 2$ for one version of CHSH, only the PR-behavior that lies on top of that same local facet contributes to the violation; as for the others, six give $S = 0$ and one $S = -4$. Obviously q_{PR} is minimized by decompositions that use only that PR-behavior: The minimal value is determined by[9]

$$S_{obs} = q_{PR}4 + (1 - q_{PR})2, \text{ i.e., } q_{PR} = \frac{S_{obs} - 2}{2}. \tag{9.6}$$

---

[8] It is still an upper bound, insofar as it is computed from the decomposition that is the most favorable for Eve. Eve may know that the actual decomposition is another one and not be able to guess so well.

[9] Contrary to the quantum case, here all that matters is the value of S_{obs}: The detailed knowledge of the behavior won't give any improvement, since every decomposition of the local part into deterministic behaviors is equivalent for Eve's guessing. In other words, randomness is uniquely determined by the "local fraction" (subsection 9.4.1). That being said, this is an accident of the (2, 2; 2, 2) scenario that does not extend to scenarios with more nonlocal extremal behaviors.

Let us then first consider the randomness of *Alice's output* for one of the inputs. For a PR-behavior, Eve's guessing probability is $\frac{1}{2}$. Thus $\bar{P} = q_{PR}\frac{1}{2} + (1 - q_{PR})$ which is

$$\bar{P} = 1 - \frac{S_{obs} - 2}{4}. \tag{9.7}$$

We observe that $P_{guess} < 1$ for all $S_{obs} > 2$: As soon as there is some nonlocality, there is randomness in a no-signaling theory. On the other hand, at the Tsirelson bound $S_{obs} = 2\sqrt{2}$ we do not have $P_{guess} = \frac{1}{2}$ as in the quantum case (8.15). This is expected: The same data certify less randomness for an adversary with more power. Turning now to the randomness of the *joint outputs* of Alice and Bob, (a_0, b_0), we notice that Eve's guessing probability for a PR-behavior is also $\frac{1}{2}$ because of the perfect correlation. Therefore, the bound on Eve's guessing probability is again (9.7). This is significantly different from the quantum case, where we have seen that there is more randomness in the joint outputs. In fact, for any Bell scenario, contrary to the quantum case (subsection 8.3.3) it is impossible to extract maximal randomness against an adversary with no-signaling description (de la Torre *et al.*, 2015).

9.3.2 Key distribution

The idea of securing key distribution (KD) against a no-signaling adversary is what triggered the awareness of device-independent certification (Appendix F.1.2). It is relatively easy to deal with an adversary with a i.i.d. no-signaling description.[10] The main difference from the previous subsection is the fact that, while randomness is a property of the behavior alone, the security of KD depends on the protocol too. A protocol based on CHSH was studied in (Acín *et al.*, 2006; Scarani *et al.*, 2006).

However, in KD we want to have a full-fledged *no-signaling adversary*, able to manipulate the corresponding side-information. Against such an adversary, it is provably impossible to define extractors along the usual lines (Arnon-Friedman and Ta-Shma, 2012; Hänggi *et al.*, 2013). Maybe unconventional constructions could lead to more optimistic results, but most of the community seems to take these results as indication that security is actually impossible.[11]

9.4 Refinements on Quantum Indeterminacy

We know since chapter 1 that, barring the adoption of higher stances leaning towards full determinism, Bell nonlocality implies intrinsic indetermination if signaling is not allowed; we have just been reminded of it in the certification of randomness. Here we are going

---

[10] This is the analog of proving security against *individual attacks* in QKD.

[11] Also, one should not forget that we are trying to base a security proof on the knowledge of the behaviors alone, and not on the underlying theory: To the best of my knowledge, it is not even clear that security against a quantum adversary could be proved in these conditions.

to see *more refined features of quantum indeterminacy* can be recovered in the no-signaling framework.

Concretely, in a classical theory, the state of a composite system is pure if and only if the state of each component is pure. This is famously false in quantum theory. We are going to see that two aspects of this statement are not just formal consequences of quantum theory, but necessities of any no-signaling theory that contains the singlet behavior (9.4).

9.4.1 The notion of local fraction

In a paper that is often referred to as EPR2, Elitzur, Popescu, and Rohrlich (1992) asked whether a source of entangled pairs may be actually two ensembles of pairs, one of which is made of classically correlated objects. This physically-posed question has a natural counterpart in the setting of games and simulations: If a behavior $\mathcal{P}$ has a local component, the players will have to worry only about a fraction of the rounds.

Formally, any no-signaling behavior $\mathcal{P}$ can be decomposed as the convex sum

$$\mathcal{P} = p\mathcal{P}_L + (1-p)\mathcal{P}_{NS} \tag{9.8}$$

where $p \in [0, 1]$ and $\mathcal{P}_L$ is local. The behavior $\mathcal{P}_{NS} \equiv \frac{\mathcal{P} - p\mathcal{P}_L}{1-p}$ is only requested to be a valid probability distribution; it is no-signaling by construction, since both $\mathcal{P}$ and $\mathcal{P}_L$ are. The *local fraction* p_L is defined as

$$p_L = \max_{(9.8)\text{ holds}} p. \tag{9.9}$$

Clearly, $\mathcal{P}$ is nonlocal if and only if $p_L < 1$. In fact, the violation $I(\mathcal{P})$ of any Bell inequality puts an upper bound on p_L (Barrett *et al.*, 2006). Indeed, from (9.8) it follows immediately that $I(\mathcal{P}) \le p I_L + (1-p)I_{NS}$, whence

$$p_L \le \frac{I_{NS} - I(\mathcal{P})}{I_{NS} - I_L}. \tag{9.10}$$

For behaviors in finite Bell scenarios, p_L can be computed algorithmically by testing all the Bell inequalities, and the corresponding $\mathcal{P}_L$ will be found as the point on a facet of the local polytope that is nearest to $\mathcal{P}$. For instance, the quantum behavior that maximizes CHSH has local fraction $p_L = 2 - \sqrt{2} \approx 0.59$.

One can also ask what the *local fraction of a state* is by considering the state behavior. At the moment of writing, no algorithmic approach has been found for this problem. One has first to conjecture that a value of p_L, maybe an upper bound (9.10), is optimal; then guess a form of the local behavior $\mathcal{P}_L$. With this, one checks if the derived $\mathcal{P}_{NS}$ is a valid probability distribution (Exercise 9.2). This approach has been successfully carried out only for the family of pure entangled states of two qubits (3.15) and for projective measurements, for which the local fraction is $p_L = 1 - \cos 2\theta$ (Portmann *et al.*, 2012).

Unfortunately, the construction of the local behavior $\mathcal{P}_L$ is an awkward tour de force, hardly inspiring generalization. As we mentioned in subsection 3.3.1, it remains an open problem to determine exactly the locality of the Werner behaviors (3.24). Beyond the case of qubits, the local fraction is virtually unexplored (Scarani, 2008).

Nonetheless, the case of the singlet is remarkable, as already proved in the EPR2 paper. For the singlet behavior (9.4), the chained inequalities lead to $I(\mathcal{P}) = I_{NS}$ in the limit of large M (subsection 4.2.3), hence $p_L = 0$. In other words, *the singlet behavior is fully nonlocal.*[12]

9.4.2   The randomness of the marginal distributions

The observation that the singlet behavior is fully nonlocal does not imply that it is extremal in the no-signaling polytope: As we have seen in section 9.2, it can be decomposed on a family of behaviors described by a PR-box complemented with local variables. That decomposition is such that

$$P_\lambda(a|\hat{a}) = P_\lambda(b|\hat{b}) = \frac{1}{2} : \qquad (9.11)$$

Even knowing λ, the local statistics are maximally random. This is analogous to the situation in quantum theory, where the partial states of maximally entangled states are maximally mixed. But is this a necessity? Can't one decompose the singlet behavior on a family of no-signaling behaviors with *biased marginals*? Without denying indetermination, such a model would at least recover some classical flavor.

The question was first posed by Tony Leggett (2003). Assuming $P_\lambda(a|\hat{a}) = \frac{1}{2}(1 + \hat{u}_\lambda \cdot \hat{a})$ and $P_\lambda(b|\hat{b}) = \frac{1}{2}(1 + \hat{v}_\lambda \cdot \hat{b})$ with $||\hat{u}_\lambda|| = ||\hat{v}_\lambda|| = 1$, that is maximal bias with the standard dependence on the measurement direction,[13] he derived an inequality that is violated by the singlet behavior. Leggett's original inequality uses an infinite quantity of inputs, but another can be found with finitely many inputs, suitable for experimental verification (it is worth noting that tests are not device-independent). These developments generated some excitement at some point, see (Branciard et al., 2008) and references therein. Eventually, the study of Leggett-type models was superseded by a device-independent proof that every no-signaling decomposition of the singlet behavior must satisfy (9.11) (Colbeck and Renner, 2008).

The proof uses the *variational distance* between two probability distributions on the same alphabet $\mathcal{C}$, defined as[14]

---

[12] In fact, any sub-behavior including all the input vectors lying in a plane is already fully nonlocal.

[13] In particular, this model is no-signaling, as each player's marginal does not depend on the other player's input.

[14] For readers familiar with quantum theory, it is the classical analog of the trace distance for states, and has the same operational interpretation as "guessing advantage."

$$\mathcal{D}[P_C, Q_C] = \frac{1}{2} \sum_{c \in C} |P_C(c) - Q_C(c)|. \tag{9.12}$$

The same definition carries over immediately to conditional probabilities, which we shall denote $P_{C|\Omega}$. The properties of the variational distance used in this proof are proved in Appendix G.7.

The first property we are going to use is the following: If the two distributions are marginals of a joint distribution with identical alphabets $\mathcal{A} = \mathcal{B}$, then

$$D(P_A, P_B) \le P_{AB}(a \ne b). \tag{9.13}$$

Let us then start from the chained inequality in its form (4.7) applied to one of the $\mathcal{P}_\lambda$. Using (9.13), we have

$$C'_M(\mathcal{P}_\lambda) \ge D(P_{A|1,1,\lambda}, P_{B|1,1,\lambda}) + D(P_{B|2,1,\lambda}, P_{A|2,1,\lambda}) + D(P_{A|2,2,\lambda}, P_{B|2,2,\lambda}) + \dots$$
$$\dots + D(P_{A|M,M,\lambda}, P_{B|M,M,\lambda}) + D(P_{B|1,M\lambda}, 1 - P_{A|1,M\lambda})$$

with the notation $P_{A|x,y,\lambda}$ and analog for Bob. No-signaling implies $P_{A|x,y,\lambda} = P_{A|x,\lambda}$ and $P_{B|x,y,\lambda} = P_{B|y,\lambda}$. Having made this replacement, we can use the fact that the distance satisfies the triangle inequality $\mathcal{D}[P, Q] + \mathcal{D}[Q, R] \ge \mathcal{D}[P, R]$ to obtain for instance

$$D(P_{A|x=1,\lambda}, P_{B|y=1,\lambda}) + D(P_{B|y=1,\lambda}, P_{A|x=2,\lambda}) \ge D(P_{A|x=1,\lambda}, P_{A|x=2,\lambda}).$$

By iterating the triangle inequality we eventually reach

$$C'_M(\mathcal{P}_\lambda) \ge D(P_{A|x=1,\lambda}, \tilde{P}_{A|x=1,\lambda}) \tag{9.14}$$

where $\tilde{P}_{A|x,\lambda}$ is defined by $\tilde{P}_\lambda(a|x) = 1 - P_\lambda(a|x)$. The r.h.s. can be computed explicitly and we finally find

$$C'_M(\mathcal{P}_\lambda) \ge \left| P_\lambda(a = 0|x = 1) - \frac{1}{2} \right| + \left| P_\lambda(a = 1|x = 1) - \frac{1}{2} \right|. \tag{9.15}$$

At this point, recall that for the singlet behavior (9.4) it holds $C'_\infty(\mathcal{P}) = 0$. Since this value is the minimum in the no-signaling set, if $C'_M(\mathcal{P}) = \int d\lambda Q(\lambda) C'_M(\mathcal{P}_\lambda)$ it must be the case that $C'_\infty(\mathcal{P}_\lambda) = 0$ for all λ. Plugging this in (9.15) forces $P_\lambda(a = 0|x = 1) = P_\lambda(a = 1|x = 1) = \frac{1}{2}$. The proof for all other inputs $x = 2, \dots, M$ and for Bob is exactly the same: One just has to keep the suitable terms when iterating the triangle inequality. Finally, to prove (9.11) for any two $\hat{a}$ and $\hat{b}$, one has just to test the chained inequality with inputs in the plane that contains those two vectors.

....

EXERCISES

Exercise 9.1. *Consider a hypothetical three-partite resource, such that the marginals Alice-Bob and Alice-Charlie are the PR-behavior: $a \oplus b = xy$ and $a \oplus c = xz$:*

(a) *Prove that such a three-partite resource is signaling. Hint: Suppose that Bob and Charlie get together: What is $b \oplus c$?*

(b) *Consider adding noise to the PR-behaviors: Now, $a \oplus b = xy$ holds with probability $1 - q$, while with probability q we have $a \oplus b = xy \oplus 1$, still with unbiased marginals; and the same for Alice-Charlie. What is the minimal value of q such that the three-partite behavior is no-signaling?*

Exercise 9.2. *Take the state behavior (3.29) associated to projective measurements on a pure, non-maximally entangled two-qubit state. Prove that the EPR2 decomposition (9.8) is valid with the guess*

$$P_L(\alpha, \beta | \hat{a}, \hat{b}) = \frac{1}{4}[1 + \alpha f(a_{\hat{z}})][1 + \beta f(b_{\hat{z}})] \quad with$$

$$f(u) = sign(u) \min\left(1, \frac{c}{1-s}|u|\right)$$

as long as $p_L \leq 1 - s$ (Scarani, 2008). Remark: It was proved that this is the highest p_L achievable with a product $\mathcal{P}_L$.

10

The Quest for Device-Independent Quantum Principles

Considerate la vostra semenza: fatti non foste a viver come bruti, ma per seguir virtute e canoscenza.

Consider your origin; you were not born to live like brutes, but to follow virtue and knowledge.

Dante, *Inferno*

No-signaling defines a larger set of behaviors than the quantum set: The additional constraints that define the quantum set could then be taken as physical principles satisfied in our universe (insofar as we know). This chapter reviews the attempts to identify such principles.

10.1 Context: The Definition of Quantum Theory

Quantum theory appeared in 1926 from the works of Schrödinger and Heisenberg, after two decades of attempts to go beyond classical physics. By the end of that same year, Born and Jordan had stressed that the two approaches are actually two versions of the same formalism. That formalizm is the one we still use today. It is widely accepted that the core difference between classical and quantum theories lies in the *description* of a physical system, whereas the axioms about evolution are basically the same. In classical theory, the space of states is a set: Physical properties are associated to subsets, pure states are the minimal subsets i.e., points. In quantum theory, the space of state is a vector space: Physical properties are associated to subspaces, pure states are the minimal subspaces i.e., one-dimensional ones or rays (more in Appendix C.1.1). This is, in essence, the *definition* of quantum theory. Since very early on, people have tried to "make sense" of this definition. This endeavour bears of course the marks of some subjectivity: There is no strict consensus about what "makes sense."

The major attempt in this direction is the program of *quantum logic*, initiated by von Neumann. The goal of the program is to *reconstruct quantum theory*: More precisely, to single it out by adding suitable axioms in a very broad class of theories. Interestingly, this program managed to find a definition of classical theory: The space of states is a set if, for every physical property, the question "does the system in this state possess

Bell Nonlocality. Valerio Scarani. © Valerio Scarani 2019. Published in 2019 by Oxford University Press.
DOI: 10.1093/oso/9780198788416.001.0001

this property?" is answered by either yes or no with certainty. Clearly our daily intuition requests this, and just as clearly quantum theory denies it. The long sought definition of quantum theory, however, took longer to be found. A breakthrough was made when, inspired by quantum information science, people started adding axioms that refer to the description of composite systems. Currently, finite-dimensional quantum theory can be fully reconstructed within the framework of "generalized probabilistic theories" (GPTs) through several set of axioms, all of which have an operational flavor (as to whether this is enough to "make sense," it's each one's call). The recent book by D'Ariano, Chiribella, and Perinotti (2017) starts with a very informative review of all this history and proceeds to present one of these reconstructions.

In 1994, Popescu and Rohrlich asked whether the principle that define quantum theory could possibly be *Bell nonlocality without signaling* (Popescu and Rohrlich, 1994). The PR-behavior was their counterexample. With the passing of the years, it was realized that their approach set the quest for "making sense" of quantum theory in a different direction: Instead of singling out quantum theory in a class of theories, one could try and *single out quantum behaviors in the no-signaling set* by adding further *principles* to that of no-signaling. None of the proposed principles actually fulfils the goal perfectly. We are going to study the three main proposals:[1] Information Causality (section 10.2), Macroscopic Locality (section 10.3) and Local Orthogonality (section 10.4).

Before delving into these, let me stress that the two approaches are complementary. On the one hand, working only with behaviors, one can define principles that are device-independent; whereas the construction of a GPT assumes that one knows the dimensionality of the system under study, in the form of fiducial sets of data that contain all the information available on the system. On the other hand, identifying a set of behaviors does not yet identify the underlying theory, possibly theories, that lead to that set.[2] One could try and merge the two approaches: There are some very preliminary results in this direction, too technical to be reviewed in this book.

10.2   Information Causality

10.2.1   The basic task: Random access code

Like van Dam's insight with communication complexity (subsection 9.2.3), Information Causality is defined through a communication task assisted by no-signaling resources. The task is the so-called *random access code* (Figure 10.1, top). We consider the simplest example. Alice's input consists of a pair of bits $(x_0, x_1) \in \{0, 1\}^2$ drawn uniformly at

---

[1]   Their names sounds like titles of episodes of the sitcom *The Big Bang Theory*. I am not aware of any causal link, beside the fact that the sitcom was very accurate in its depiction of academia and its jargon.

[2]   At the cost of repeating ourselves: The approach through behaviors does *not* aim at singling out "quantum theory among no-signaling theories," as casually stated in many papers on the subject.

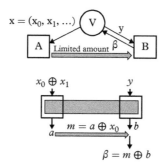

Figure 10.1 *Top: The task that inspires Information Causality is a delocalized random access code: Bob's output β is supposed to be equal to x_y, the element of Alice's string indexed by Bob's input. The task is made non-trivial by allowing a limited amount of communication from Alice to Bob. Bottom: With the use of a PR-box (a no-signaling resource), Bob can successfully output $β = x_y$ in every round, with Alice sending him only one bit of communication m.*

random among the four possible values. Bob's input is a bit $y \in \{0, 1\}$ unknown to Alice *a priori*.[3] Bob is asked to output x_y.

If Alice cannot send any information to Bob, Bob's output will obviously be uncorrelated with Alice's inputs. If Alice can send two bits, Bob's output can be correlated as desired. The interesting case is when Alice is *restricted to send only one bit*. Without shared resources, it's clear what is the best strategy: Alice sends always one of her bits, say x_0; Bob outputs the bit he receives: If $y = 0$, his output is the desired one, while if $y = 1$ his output is uncorrelated with the desired answer x_1.

But could Alice and Bob be more successful if they would share no-signaling resources? Intuition suggests a negative answer: The piece of information that Bob needs is with Alice, it can't arrive to him without communication, so a no-signaling resource should not help. However, this intuition is wrong: If Alice and Bob could share a PR-box, they could win every round of the game!

10.2.2 The power of PR-boxes, again

Here goes the protocol (Figure 10.1, bottom): Alice inputs $x = x_0 \oplus x_1$ in the PR-box; she gets the output a and sends to Bob the single-bit message $m = a \oplus x_0$. Bob inputs y in the PR-box, gets the output b and outputs $β = m \oplus b = (a \oplus b) \oplus x_0$. By the rule (9.1) of the PR-box, $a \oplus b = (x_0 \oplus x_1)y$; so $β = x_0 \oplus [(x_0 \oplus x_1)y] = x_y$ as claimed.

Notice that Bob had to put y into the PR-box to retrieve b: Consequently, he gains no information on the other bit x_{1-y}. In this sense, nothing has gone blatantly wrong: Alice sent one bit and Bob got one bit. The PR-box has not increased the amount of information transferred between the players, which is as it should be since it is a

---

[3] If one adds the requirement that Alice is forbidden to know Bob's choice even *a posteriori*, the task is known as *oblivious transfer*. We don't enforce this requirement here.

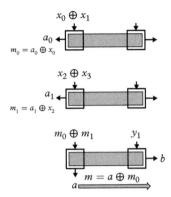

Figure 10.2 *Nested protocol to retrieve one out of $N = 4$ bits (x_0, x_1, x_2, x_3) with one bit of communication, that works perfectly for the PR-box. Referring to the basic protocol sketched in Figure 10.1 bottom, we see that m_0 (respectively, m_1) is what Alice should send to Bob to retrieve either x_0 or x_1 (respectively, either x_2 or x_3). Using the third box and the basic protocol, Bob can retrieve either m_0 or m_1: With this piece of information, he can then retrieve the desired x_y querying either the first or the second box (this last step is not shown).*

no-signaling resource. It is nevertheless puzzling that Bob can retrieve perfectly whichever bit he chooses to read out: It looks as if both bits had been transferred to his location, even if ultimately he is allowed to access only one.[4] Information Causality (IC) basically states that this should not be possible (Pawłowski *et al.*, 2009).

Before formally defining IC, let us indulge further in the power of the PR-box by generalizing the random access code to more inputs. Consider $\vec{x} \in \{0, 1\}^4$ and $y \in \{0, 1, 2, 3\}$, each drawn with uniform distribution. Bob can retrieve perfectly each of the four bits if Alice communicated to him only one bit, provided they share three PR-boxes. This is done with the following nested protocol (Figure 10.2):

- Alice inputs $x_0 \oplus x_1$ in the first PR-box, and gets the output a_0; she inputs $x_2 \oplus x_3$ in the second PR-box, and gets the output a_1. Finally, Alice inputs $m_0 \oplus m_1 \equiv (a_0 \oplus x_0) \oplus (a_1 \oplus x_2)$ into the third PR-box, gets the output a. She sends the single bit $m = m_0 \oplus a$ to Bob.

- Bob has received y, which we can write in binary notation as $y = 2y_1 + y_0$ with $y_0, y_1 \in \{0, 1\}$. He inputs y_1 in the third PR-box, and upon receiving the output b he can retrieve $m_{y_1} = m \oplus b$. Now, if $y_1 = 0$ Bob must retrieve either x_0 or x_1: But he has just retrieved m_0, so he can apply the protocol of Figure 10.1 bottom by inputting y_0 to the first PR-box. Similarly, if $y_1 = 1$ Bob must retrieve either x_2 or

---

[4] Or, following a different narrative: Suppose that Alice wants to send two books to Bob, but her communication channel allows her to send only one. We expect that she would have to make a choice. But with PR-boxes, she could encode information about both books, in such a way that Bob can retrieve the one of his choice.

x_3: Having just retrieved m_1, he can apply the protocol of Figure 10.1 bottom by inputting y_0 to the second box. Notice that Bob does not need to query one of the boxes.

This construction generalizes to a random access code with $\vec{x} \in \{0, 1\}^N$ and $y \in \{0, 1, \ldots N - 1\}$ for every $N = 2^n$ and $n \in \mathbb{N}$. The nested protocol has then n layers and requires 2^{n-k} PR-boxes in the k-th layer, for a total of $N - 1$ PR-boxes. Using this protocol, with just one bit of communication from Alice, Bob can retrieve *any* of her N input bits with certainty, while learning nothing about the others. In any layer, Bob has to query only one box.

10.2.3   Definition of Information Causality

Let us denote by $\boldsymbol{\beta}$ the alphabet of Bob's output β, by $\mathbf{X}$ the alphabet of each of Alice's inputs – it's of course the same alphabet, and for all the examples here we'll have $\boldsymbol{\beta} = \mathbf{X} = \{0, 1\}$. For every y, the protocol that Alice and Bob adopt leads to a family of joint probability distributions $P_{\beta_y, \mathbf{x}_y}(\beta_y, x_y)$ for Bob's output and the input he is asked to guess. The amount of information that β_y carries about x_y is quantified by the *mutual information*

$$I(\boldsymbol{\beta}_y : \mathbf{X}_y) = H(\boldsymbol{\beta}_y) + H(\mathbf{X}_y) - H(\boldsymbol{\beta}_y, \mathbf{X}_y) \tag{10.1}$$

where $H(C) = -\sum_{c \in C} P_C(c) \log_2 P_C(c)$ is Shannon's entropy. By definition of the task, Alice's inputs are uniformly distributed, whence we'll have always $H(\mathbf{X}_y) = \log_2 |\mathbf{X}|$. When $\beta = \mathbf{X} = \{0, 1\}$, it holds

$$I(\boldsymbol{\beta}_y : \mathbf{X}_y) = 1 - h\left(P_{\beta_y, \mathbf{x}_y}(\beta_y = x_y)\right) \tag{10.2}$$

where $h(p) = -p \log_2 p - (1 - p) \log_2(1 - p)$ is the binary Shannon entropy.

Let us study two examples when Alice's input is $N = 2^n$ bits (recall that Alice's inputs are uniformly distributed by definition):

- If the players don't share any additional resources, and if Alice sends the κ bits $\vec{m} = (x_0, \ldots, x_{\kappa-1})$ to Bob, then:

  - For $y = 0, \ldots, \kappa - 1$, Bob's output is equal to Alice's desired input, i.e., $P_{\beta_y, \mathbf{x}_y}(\beta, x_y) = \frac{1}{2}\delta_{\beta_y, x_y}$. Therefore $H(\boldsymbol{\beta}_y) = H(\mathbf{X}_y) = H(\boldsymbol{\beta}_y, \mathbf{X}_y) = 1$ and finally $I(\boldsymbol{\beta}_y : \mathbf{X}_y) = 1$.
  - For $y = \kappa, \ldots, N - 1$, Bob's outcome is uncorrelated from Alice's input: $P_{\beta_y, \mathbf{x}_y}(\beta_y, x_y) = \frac{1}{2}P(\beta_y)$, where $P(\beta)$ depends on what Bob does. In any case, $H(\boldsymbol{\beta}_y, \mathbf{X}_y) = H(\boldsymbol{\beta}_y) + H(\mathbf{X}_y)$ and therefore $I(\boldsymbol{\beta}_y : \mathbf{X}_y) = 0$.

- If the players share $N - 1$ PR-boxes, we have seen that $P_{\boldsymbol{\beta}_y, \mathbf{x}_y}(\beta_y, x_y) = \frac{1}{2}\delta_{\beta_y, x_y}$ for all y as soon as Alice can send one bit of communication. So $I(\boldsymbol{\beta}_y : \mathbf{X}_y) = 1$ for all y.

Now, Information Causality is supposed to capture the "non-aberrant" behavior in a random access code, so these examples suggest the following definition (Pawłowski *et al.*, 2009): In a random access code with an input alphabet $\mathbf{X}^N$, *Information Causality is satisfied* if

$$I \equiv \sum_{y=1}^{N} I(\boldsymbol{\beta}_y : \mathbf{X}_y) \stackrel{IC}{\leq} \kappa \tag{10.3}$$

where $\kappa = |\vec{m}| \in [0, N \log_2 |\mathbf{X}|]$ is the number of bits that Alice communicates to Bob.

It can be proved that (10.3) holds if the shared no-signaling resources are described by quantum theory, which of course implies that it holds also for LVs—and we know, from the first bullet listed previously, that (10.3) holds even in the absence of shared resources.[5] In fact, IC is satisfied under very reasonable requirements on the information theory on the underlying resources. The first proof used three requirements on the mutual information (Pawłowski *et al.*, 2009); a later simplification (Al-Safi and Short, 2011) showed that IC follows from only two requirements on the entropy function H. The first is: Whenever the theory describes a resource that can be described in classical theory, the value of the entropy must coincide with the Shannon entropy. The second is a form of the data-processing inequality, that says that if a transformation is made on Bob's system, its correlations with Alice's system should not increase. The details and the proof are given in Appendix G.8.

These requirements definitely hold in quantum information theory, but we don't know if one can construct other theories where they hold. To explore whether IC singles out the quantum set, the best we can do at this stage is to work out explicit calculations based on the behaviors.

10.2.4   Recovering the Tsirelson bound with IC

The most striking success of IC is that it can recover the Tsirelson bound (3.5) for the (2, 2; 2, 2) scenario without any reference to the algebra of Hilbert spaces, something that could not be achieved building on van Dam's observation (Brassard *et al.*, 2006). Assume that Alice and Bob share *isotropic boxes* defined as hypothetical resources that produce the behavior

$$P_V(a, b|x, y) = VP_{PR}(a, b|x, y) + (1 - V)/4. \tag{10.4}$$

---

[5] This agreement between the trivial, LV and quantum bounds is a very nice feature of this definition of IC. It would not hold if, instead of using mutual information, we had defined IC using Bob's guessing probability (Al-Safi and Short, 2011).

This behavior gives $S = 4V$ for CHSH, and for $V = \frac{1}{\sqrt{2}}$ the behavior (10.4) is the same as the quantum behavior (3.11). Then, IC cannot be violated for $V \leq \frac{1}{\sqrt{2}}$, because in this case the isotropic box can be realized with quantum resources. We want to show that IC is violated as soon as $V > \frac{1}{\sqrt{2}}$.

Let's assume that Alice communicates $\kappa = 1$ bits to Bob. For $N = 2$ input bits we have $P(\beta_y = x_y) = V \times 1 + (1 - V) \times \frac{1}{2} = \frac{1+V}{2}$ for all choices of $\vec{x}$, y. So, IC will be violated if $2\left[1 - h\left(\frac{1+V}{2}\right)\right] > 1$, that is for $V \gtrsim 0.78$. This is not yet the result we are looking for. But when $N = 2^n$, the nested protocol introduced in subsection 10.2.2 applied to the isotropic box gives

$$P(\beta = x_y | \vec{x}, y) = \frac{1 + V^n}{2} \tag{10.5}$$

for all choices of $\vec{x}$, y (Exercise 10.1 for $N = 4$). Crucially, the error does not grow too fast because only n boxes are actually used: Recall that Bob uses only one box per layer. Using the Taylor expansion $1 - h\left(\frac{1+V^n}{2}\right) \geq \frac{V^{2n}}{2\ln 2}$, we find that IC is violated if

$$2^n\left[1 - h\left(\frac{1 + V^n}{2}\right)\right] \geq \frac{(2V^2)^n}{2\ln 2} > 1. \tag{10.6}$$

In other words, for every value $V > \frac{1}{\sqrt{2}}$, there exist a n such that IC is violated, which concludes the proof.

10.2.5   Information Causality as a physical principle?

Remarkable though it is, the recovery of the Tsirelson bound refers to one extremal point of the quantum set of the (2, 2; 2, 2) Bell scenario. Results for other parts of the quantum set and other Bell scenarios have been difficult to obtain. One of the difficulties is that we don't know any necessary condition for the violation of IC. For instance, for some families of boxes the nested protocol described previously does not reach the quantum set (Allcock *et al.*, 2009): In this case, we don't know if it is because the protocol is not optimal, or because indeed the remaining boxes satisfy IC. Overall, it is still possible that IC identifies the quantum set, but most people conjecture the opposite.

Furthermore, the extension of IC to multipartite Bell scenarios has not been successful: All attempts at generalization tried so far ultimately reduce to a bipartite task, and a very general argument shows that no bipartite task can single out the quantum set of multipartite Bell scenarios (Gallego *et al.*, 2011). All these developments have been described in (Pawłowski and Scarani, 2016): Contrary to the wish expressed there, research in IC has not significantly progressed since.

At any rate, it would be wrong to end on a gloomy note. Even if IC is not the desired quantum principle, it is a property of all quantum resources and therefore of nature as we currently know it. The impossibility of outperforming classical resources in that communication task adds to the no-go theorems conclusively proved in quantum information science.

10.3　Macroscopic Locality

10.3.1　Motivation and formulation

The possibility of measuring individual quantum systems is so crucial to present-day quantum science and technologies, that we may forget that it's a relatively recent development in the laboratories. Only in the 1970's it became possible to observe interference effects while recording individual events.[6] Interestingly, these are the same years when the first attempts at Bell test were being conducted, and less than a decade before the first conclusive demonstrations of Bell nonlocality (Aspect *et al.*, 1982*b*, Aspect *et al.*, 1982*a*).

One may wonder: If Aspect and coworkers would not have had single-quanta detection techniques, if they could only have measured rather intense currents of photons,[7] would they have been able to observe nonlocality? In the limit of very large currents, the answer is no, as we are going to show. Just as Popescu and Rohrlich did with no-signaling, Navascués and Wunderlich (2010) turned this observation into a candidate for the quantum principle. They called *macroscopically local* those behaviors, such that a coarse-graining over many systems washes nonlocality out; as usual, we adopt this wording without debating its appropriateness.[8]

For the formal definition of *macroscopic locality (ML)*, consider a $(M_A, m; M_B, m)$ Bell scenario, labeling the outputs as $a, b \in \{0, 1, \ldots, m - 1\}$ for convenience. The players have agreed on a i.i.d. strategy that would generate the behavior $\mathcal{P}$. However, the verifier is not going to play the usual Bell test. Rather, he sends the same inputs (x, y) for N consecutive queries; from the string of outputs $(a_x^{(1)}, \ldots, a_x^{(N)})$ and $(b_y^{(1)}, \ldots, b_y^{(N)})$ collected in one such round, he just records *the number of occurrences of each value in the string*

$$\mathbf{I}_x = (I_{0|x}, I_{1|x}, \ldots, I_{m-1|x}) \ \text{ and } \ \mathbf{J}_y = (\mathcal{J}_{0|y}, \mathcal{J}_{1|y}, \ldots, \mathcal{J}_{m-1|y}), \tag{10.7}$$

with

$$I_{a|x} = \sum_{r=1}^{N} \delta_{a_x^{(r)} = a}, \ \ \mathcal{J}_{b|y} = \sum_{r=1}^{N} \delta_{b_y^{(r)} = b}. \tag{10.8}$$

---

[6] For electron interferometry, it seems that the first demonstration with individual detections was that of Merli, Missiroli, and Pozzi at Bologna in 1974. In the same year, single-neutron interferometry was demonstrated by Helmut Rauch (Vienna and Grenoble) and Samuel A. Werner (Missouri and Michigan).

[7] Their source based on atomic cascade producing very few events in the relevant modes, it would have been very difficult to *produce* such intense currents; but this is beside the point here.

[8] Certainly macroscopic locality does not pretend to capture either the essence, or even an essential feature, of *macroscopicity*. This notion has a long history of formulations and is still the object of studies and discussions. Topics like "emergence" in condensed matter physics, or the "sorites paradox" in philosophy, are related to it. In the foundations of quantum mechanics, macroscopicity plays a double role. First, in the attempts at explaining why we typically don't see quantum effects at our everyday scale: ML builds on this, focusing on the impossibility of observing Bell nonlocality. Second, in the efforts to demonstrate quantum effects with unusually large objects ("Schrödinger cats").

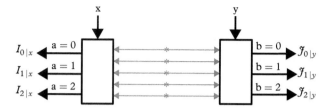

Figure 10.3 *The coarse-graining that is used in the definition of Macroscopic Locality. The resource consists of N identical sources, or N uses of the same source (N = 5 in the figure). The players record only the number of times each output is produced. By doing so, their local statistics are unchanged, but the correlations between the two players are coarse-grained.*

Translated into laboratory language, it captures the situation where a source is emitting i.i.d. pairs that are measured in groups of N at each station, and only the resulting *currents* are recorded (Figure 10.3). In this coarse-graining, what is lost is the information about which output of Alice's was correlated with which output of Bob's.

If the underlying behavior is the PR-behavior (9.3), the verifier can still certify nonlocality after the coarse-graining. Indeed, for any N, each round would exhibit $I_{0|x} = \mathcal{J}_{0|y}$ for $xy = 0$ and $I_{0|1} = \mathcal{J}_{1|1}$ for $xy = 1$; and notice that $I_{1|x} = N - I_{0|x}$ and $\mathcal{J}_{1|y} = N - \mathcal{J}_{0|y}$ since $m = 2$. Suppose now that one applies the local post-processing of *majority vote*: Alice outputs $\alpha = 0$ if $I_{0|x} \geq \frac{N}{2}$ and $\alpha = 1$ if not; and Bob does the same for β. The post-processed behavior $P(\alpha, \beta|x, y)$ thus obtained is the PR-behavior itself.[9] Even before a majority vote, therefore, the coarse-grained data contained nonlocality.

10.3.2   The set of ML behaviors is $\mathcal{Q}_1$

The *principle of macroscopic locality* states that the verifier *should not* be able to certify nonlocality after the coarse-graining, in the limit $N \to \infty$. We have just seen that the principle rules out the PR-behavior. Following (Navascués and Wunderlich, 2010), we are going to see that *the set of ML behaviors is identical to* $\mathcal{Q}_1$, the first step of the NPA hierarchy.

The proof combines two observations. First, by Fine's theorem, the behavior $P(\mathbf{I}_x, \mathbf{J}_y)$ of the coarse-grained currents is local if and only if it is the marginal of a joint probability distribution $\mathbf{P}(\mathbf{I}_1, \mathbf{I}_2, \ldots, \mathbf{I}_{M_A}; \mathbf{J}_1, \mathbf{J}_2, \ldots, \mathbf{J}_{M_B})$. Second, the individual currents (10.8) are sums of i.i.d. random variables; as a consequence, each $\mathbf{I}_x$ and each $\mathbf{J}_y$ are sums of i.i.d. random vectors. Both observations are valid for all N, but in the limit $N \to \infty$ the *central limit theorem* provides a very powerful characterization of ML: If $\mathbf{P}$ exists, and if the detectors have finite resolution of the order $O(\sqrt{N})$, the fluctuations of its variables around their average must obey a multivariate Gaussian distribution with zero average

---

[9] More precisely, rounds for which $\mathcal{J}_{1|1} = \mathcal{J}_{0|1} = \frac{N}{2}$ exactly would lead to $\alpha = \beta = 0$ in spite of $xy = 1$. But such rounds are impossible if N is odd, and of negligible probability if N is even and large.

and covariance matrix $\Gamma \geq 0$. Inversely, if such a Gaussian distribution exists, it defines a valid $\mathbf{P}$.

The proof will be finished by showing that Γ is the matrix that defines $\mathcal{Q}_1$ for the single-round behavior $\mathcal{P}$. Indeed, let's define the fluctuations as

$$f_{a|x} = \frac{I_{a|x} - \langle I_{a|x} \rangle}{\sqrt{N}}, f_{b|y} = \frac{\mathcal{J}_{b|y} - \langle \mathcal{J}_{b|y} \rangle}{\sqrt{N}} : \tag{10.9}$$

the entries of the covariance matrix are the $\Gamma_{ij} = \langle \mathbf{f}_i \cdot \mathbf{f}_j \rangle$, i.e., with a suitable labeling

$$\Gamma = \left(\frac{\{\langle \mathbf{f}_x \cdot \mathbf{f}_{x'} \rangle\} \mid \{\langle \mathbf{f}_x \cdot \mathbf{f}_{y'} \rangle\}}{\{\langle \mathbf{f}_y \cdot \mathbf{f}_{x'} \rangle\} \mid \{\langle \mathbf{f}_y \cdot \mathbf{f}_{y'} \rangle\}} \right). \tag{10.10}$$

The elements of the off-diagonal blocks are terms of the form

$$\langle f_{a|x} f_{b|y} \rangle = \frac{1}{N} (\langle I_{a|x} \mathcal{J}_{b|y} \rangle - \langle I_{a|x} \rangle \langle \mathcal{J}_{b|y} \rangle)$$

$$\xrightarrow{N \to \infty} P(a, b|x, y) - P(a|x)P(b|y). \tag{10.11}$$

The terms of the diagonal blocks have a similar form, but can be associated to observable probabilities only when $x = x'$ ($y = y'$). The matrix (10.10) defines indeed the step $\mathcal{Q}_1$ of the NPA hierarchy, up to redefining the measurement operators as $F_{a|x} = \Pi_a^x - \langle \Pi_a^x \rangle \mathbb{I}$ and $F_{b|y} = \Pi_b^y - \langle \Pi_b^y \rangle \mathbb{I}$.

As a corollary of this proof, notice that, under the *promise* that the players are using an i.i.d. strategy, the verifier can reconstruct $\mathcal{P}$ from the covariance matrix of the coarse-grained statistics. Thus, he may infer that they share a nonlocal resource, whose nonlocality could be manifest if fine-grained detection would be available.

10.3.3 Summary of Macroscopic Locality

Just as with IC, all quantum behaviors satisfy ML: The intuition that one should not be able to observe nonlocality in those conditions is vindicated. As a candidate for the quantum principle, ML definitely does not single out the quantum set: Rather, it provides a physical definition of the larger set $\mathcal{Q}_1$ in any Bell scenario, including multipartite ones (both the definition and the proof extend naturally). In particular, like IC, the Tsirelson bound is recovered in the (2, 2; 2, 2) scenario. There are behaviors that satisfy ML but violate IC (Cavalcanti *et al.*, 2010).

Can one come closer to $\mathcal{Q}$ with similar ideas? One could study ML for finite N: This certainly defines a smaller set of behaviors, because the nonlocality of some $\mathcal{P}$ will not be washed out. Also, one may consider unrealistic resolutions of the detectors, so that other moments of the limiting distribution must be studied. Initial studies in these directions uncovered features that may be of interest for the experts, but seem to indicate that one does not recover Q in any case (Zhou *et al.*, 2017).

10.4   Local Orthogonality, a.k.a. Consistent Exclusivity

The idea of *local orthogonality* (LO) is based on a simple observation. Two probabilities $P(a, \ldots | x, \ldots)$ and $P(a', \ldots | x, \ldots)$ are associated in quantum theory to orthogonal projectors as soon as $a \neq a'$, whatever the inputs and outputs of the other players. As a consequence, in quantum theory one has $P(a, \ldots | x, \ldots) + P(a', \ldots | x, \ldots) \leq 1$. This idea extends to sums of more terms; in fact, we have already used it to prove the quantum bound of the GYNI inequality (5.3). But GYNI is violated by no-signaling behaviors: So this restriction can be erected as a principle that may single quantum theory out. Explicitly, *the principle of local orthogonality* requests

$$\sum_{a,b,c\ldots;x,y,z,\ldots} P(a, b, c, \ldots | x, y, z, \ldots) \leq 1 \qquad (10.12)$$

if, for every pair of terms in the sum, there is at least one player that has the same input but different output (Fritz *et al.*, 2013). Almost simultaneously, the analogos mathematical idea was being proposed as a principle for contextuality and called *exclusivity principle* (Cabello, 2013). The wording *consistent exclusivity* was used in a graph-theoretical synthesis of all relevant notions and results for both nonlocality and contextuality (Acín *et al.*, 2015).

For bipartite Bell scenarios, LO is automatically satisfied by all no-signaling behaviors, as first proved in (Cabello *et al.*, 2010). However, some hypothetical bipartite no-signaling resources can be ruled out by LO by using them in multipartite Bell scenarios. For instance, the behavior describing two independent PR-boxes $P(a, b, c, d|x, y, z, w) = P_{PR}(a, b|x, y)P_{PR}(c, d|z, w)$ violate LO in a four-partite scenario since it gives

$$P(0, 0, 0, 0|0, 0, 0, 0) + P(1, 1, 1, 0|0, 0, 1, 1) + P(0, 0, 1, 1|0, 1, 1, 0)$$

$$+P(1, 1, 0, 1|1, 0, 1, 1) + P(0, 1, 1, 1|1, 1, 0, 1) = \frac{5}{4} \quad (10.13)$$

whereas LO requires this sum of probabilities to be bounded by 1 (see also Exercise 10.2).

Thus, a behavior may satisfy LO for one copy but not for two (or more) independent copies, which makes the study of LO rather convoluted. For instance, it was proved that the set of behaviors $\mathcal{P}$ such that n independent copies satisfy LO is not convex even in the limit $n \to \infty$. One can also consider independent copies of *different* behaviors. In this context, consider the composition of a behavior $\mathcal{P}$ with an arbitrary quantum behavior $\mathcal{P}_Q \in \mathcal{Q}$: Then $\mathcal{P} \times \mathcal{P}'$ satisfies LO if and only if $\mathcal{P}$ belongs to the "almost-quantum set" described in the next section; and this remains the case also if one consider the composition of multiple copies. For the proofs of these results and others, we refer to[10] (Acín *et al.*, 2015). But in summary, also the LO principle fails to identify the quantum set.

10.5   The Current Barrier: The "Almost-Quantum" Set

Even taken together, the three principles of Information Causality, Macroscopic Locality and Local Orthogonality do not single out the quantum set. Rather than proposing other principles, some authors suggested that one cannot do much better (Navascués *et al.*, 2015). Their observation is a powerful one, although it does not have the full strength of a no-go theorem.

The definition of $\mathcal{Q}'$ is based on the requirement (6.2) that the operators that describe measurements by different players must all commute. The idea is to relax this requirement by demanding that those operators commute *when acting on the state* that is used to generate behavior:

$$[\Pi_a^x, \Pi_b^y]\,|\psi\rangle = 0 \tag{10.14}$$

for $P(a,b|x,y) = \langle\psi|\Pi_a^x\Pi_b^y|\psi\rangle$. Notice that, for this formulation, it is convenient to make full use of the unspecified Hilbert space dimensions and assume both the state to be pure and the measurements to be projective. The generalization for multipartite scenarios is obvious. The set defined by this condition was called the *"almost-quantum"* set. It is *strictly larger* than $\mathcal{Q}'$: For bipartite scenarios, it is $\mathcal{Q}_{1+AB}$ in the NPA hierarchy. It was then proved that all behaviors in the almost-quantum set satisfy both ML and LO. At the moment of writing, it has not been proved that they satisfy IC as well, but all the extant examples are compatible with such a conjecture.

More importantly, it has been conjectured that no criterion defined only on observed behaviors will come closer to $\mathcal{Q}'$ than the almost-quantum set. The basis for the conjecture is the observation that a behavior carries information about the action of the operators *only* on the corresponding state. While this is not a tight mathematical proof, it is a very convincing argument. Current attempts of coming closer to the quantum set complement principles defined on behaviors with structural properties of the underlying theory.

---

[10] It must be noted that Q_1 in that reference is not $\mathcal{Q}_1$ of the NPA hierarchy as defined in this book: It is a definition that is convenient both for contextuality and nonlocality, and it does correspond to the "almost-quantum" set in the latter case.

....

EXERCISES

Exercise 10.1. *We consider the nested protocol sketched in Figure 10.2:*

1. *For $N = 4$ input bits to Alice, prove that the protocol gives $P(\beta = x_y | \vec{x}, y) = \frac{1+V^2}{2}$ for the isotropic boxes defined by* (10.4).
2. *Sketch the drawing of the protocol in the case of $N = 8$ input bits to Alice.*

Exercise 10.2. *Find the value of the expression on the left-hand side of* (10.13) *when $P(a, b, c, d | x, y, z, w) = P_V(a, b | x, y) P_V(c, d | z, w)$, a product of two isotropic boxes* (10.4). *Until which value of V is the principle of Local Orthogonality violated?*

11

Signaling and Measurement Dependence

When you have eliminated the impossible, whatever remains, however improbable, must be the truth.

A. Conan Doyle, *The Sign of Four*

Communication (signaling) and measurement dependence are the two possible mechanisms, by which information about some players' inputs may be available at other players' location. In this last chapter, we study models that allow for such channels, relaxing the strict requirements of no-signaling and measurement independence.

11.1 Motivation: Towards Ultimate Relaxations

After some discussions in the second half of chapter 1, we have focused on describing nonlocality for *no-signaling resources* under the assumption of *measurement independence*. Besides being in tune with the most common interpretations of quantum theory and of our scientific activity in general, these two assumptions are necessary in the following sense: With unrestricted communication, or with unrestricted correlations between the shared resource and the inputs, the players can pass trivially any Bell test.

However, we are going to see that Bell nonlocality is robust to *some relaxations* of those assumptions. This can be understood: Both mechanisms are ultimately a way of carrying information about one player's input to other players: If the information is not sufficient, the Bell test retains some relevance.

Admittedly, once a communication channel is open, or once correlations between resource and inputs are allowed, there are no obvious reasons to restrict the shared information. This is the reason why these topics have not been studied thoroughly: Most experts seem to be content with knowing that no-signaling and measurement independence are not tight straitjackets. Besides, only preliminary studies have attempted to present both relaxations in the same framework (Hall, 2011; Barrett and Gisin, 2011; Chaves, *et al.*, 2015).

We shall discuss the relaxation of no-signaling in section 11.2, that of measurement independence in section 11.4. Between the two, section 11.3 is devoted to another

Bell Nonlocality. Valerio Scarani. © Valerio Scarani 2019. Published in 2019 by Oxford University Press.
DOI: 10.1093/oso/9780198788416.001.0001

aspect of communication, namely the speed at which such hypothetical influences should propagate. Finally, section 11.5 presents a new take on randomness in the context of measurement dependence.

11.2   Signaling Models: The Information in the Communication

We begin by studying models that would produce nonlocal behaviors using *actual communication*, in short *signaling models*. We shall use the word "influences," rather than "signals," for these hypothetical mechanisms of transferring information.

There are at least two reasons to be interested in these models. First, they are one of the few ways to recover determinism: It's good to have a clearer idea of the price one has to pay. Second, communication being a process that we are familiar with, they can provide an insight on the power of Bell nonlocality.

In this section, we look at the *information* that the communication must convey, leaving aside the infamous fact that these influences should propagate faster than the speed of light. The conflict between such models and relativistic space-time descriptions is addressed in the next section.

11.2.1   Fine tuning

When considering the information carried by the influences in a signaling model, we must start by pointing out that these models have a highly conspiratorial flavor: If nature is using communication to produce quantum behaviors, why is it hiding it? To my knowledge, nobody has ever attempted to propose an answer.

For those who think that "why" questions should not be asked in physics, the conspiracy can be cast in a different way: In order to simulate no-signaling behaviors with communication, the use of the communication channel should be very finely tuned. Any deviation from this *fine tuning* would produce behaviors that violate the no-signaling conditions,[1] thus uncovering the underlying mechanism. Confirming what one would intuit, every signaling model must indeed be finely tuned if it has to simulate quantum predictions (Wood and Spekkens, 2015; Cavalcanti, 2018).

11.2.2   Signaling in Bell scenarios

The influence needs only to carry information about the inputs. In particular, in a Bell scenario $(M_A, m_A; M_B, m_B)$, all no-signaling behaviors can be trivially realized if Alice can send to Bob a M_A-valued symbol per round (i.e., $\log_2 M_A$ bits), or Bob send to Alice a M_B-valued symbol ($\log_2 M_B$ bits). Some behaviors may be simulated with much less

---

[1] Insofar as we can tell, this fine tuning is *not* related to that of the parameters of the universe, which inspires the discussions on the "anthropic principle."

communication: This is what is studied in the field of communication complexity. The review (Buhrman *et al.*, 2010) is devoted to its connection with nonlocality.

An arbitrary small amount of signaling may be sufficient to produce nonlocality, even if it is hidden in a no-signaling behavior, as illustrated by the following protocol. In each round, Alice and Bob share either $(a_0, a_1, b_0, b_1) = (+1, +1, +1, +1)$ or $(-1, -1, -1, -1)$ with same probability. If Alice receives the input $x = 1$, with probability p_f she sends this information to Bob, who would then flip his bit if he received $y = 1$. The resulting behavior is $P(a,b|x,y) = \frac{1}{2}[1 + ab(1 - 2p_f xy)]$: It satisfies the no-signaling condition, since $P(a|x) = P(b|y) = \frac{1}{2}$, and violates CHSH for any $p_f > 0$, since $S = 2(1 + p_f)$.

11.2.3   Simulating state-behaviors with communication

If Alice can choose an arbitrary measurement, her output will steer a corresponding state on Bob's side: Since there are continuously many possibilities, one then may naively guess that the simulation of a state-behavior requires an infinite amount of communication. This is certainly the case if the players do not share any correlation; but if the players can share LVs, a finite amount may be sufficient—and, as it turns out, it *is* sufficient for all the known examples of simulations of state-behaviors [see section VI.C of (Buhrman *et al.*, 2010)]. Besides, the amount of communication *averaged over the measurements* (including POVMs) is bounded by $O(m^2)$ for any state-behavior with arbitrarily many players and m outputs per player (Brassard *et al.*, 2019).

Concretely, the singlet behavior (9.4) can be simulated with *a single bit of communication* in each round (Toner and Bacon, 2003). In the version of (Degorre *et al.*, 2005), the model is reminiscent of Werner's construction of a LV model (subsection 3.3.1). Alice and Bob share two vectors drawn uniformly on the unit sphere. Alice selects the vector that has the largest overlap (in absolute value) with her measurement direction and communicates this piece of information to Bob; their output is determined by the sign of the scalar product between that vector and their measurement direction. The explicit calculation is given in Appendix G.6. These protocols are explicit examples of the fine tuning mentioned in subsection 11.2.1: If they were taken as an accurate description of what happens in nature, there is no clear reason why those resources should be used in such a contrived way.

At any rate, having found a solution for the paradigmatic case of the singlet behavior, one could have hoped for quick extensions and generalizations. There have been indeed some extensions (Degorre *et al.*, 2009). But there have also been frustrating difficulties: At the moment of writing, it is not even known if the state behavior associated to a pure, non-maximally entangled state of two qubits can be simulated with a single bit of communication.[2] We also notice the connection with the (equally frustrating) study of

---

[2] The trouble seems to be associated with the "Hardy ladder" that we encountered in subsection 4.4.1: Most measurements of Alice steer a set of non-orthogonal states on Bob's side. This observation leads to a rather inelegant simulation that uses *two* bits of communication, and goes as follows. First, Alice chooses her measurement direction $\hat{a}$ and generates her output by sampling from the distribution of the partial state. With this knowledge, she knows which state she must steer on Bob's side, call it $|+\hat{n}\rangle$. Now, she turns to the

the local fraction presented in subsection 9.4.1: Communication would only be needed in a fraction of rounds, but one must still find how much is needed to reproduce the corresponding nonlocal behavior.

11.3   Signaling Models: The Speed of the Influence

11.3.1   The trouble with "faster than light"

I want to start this section by reminding the status of the principle of relativity. It is trivial that one should be able to transform one's own coordinates onto those in which the phenomenon under study is going to be described. Galileo's principle of relativity postulates that this transformation is never compulsory (though it is very handy and routinely used): The laws of physics should be the same in all (inertial) frames. When applied to electromagnetism, this principle leads to the Lorentz group of transformations, and nicely agree with the observation that the speed of light in vacuum c is independent of the observer's state of motion. Having the Lorentz group, it is a classic exercise to prove that, if we were able to send signals faster than light, we could send information to our past. To do so, observer A sends a signal at a speed v seen in *her* Lorentz rest frame. Observer B, who is moving away from A, replies with a signal at a speed v seen in *his* Lorentz rest frame. A simple space-time diagram proves that A receives B's reply before sending out her query if $v > c$. Since sending information to the past is regarded as absurd, no signal should propagate faster than light in vacuum.[3]

The most elementary signaling model for Bell nonlocality assumes that there is a *preferred frame* for the *influence* that carries the information. In this frame, one of the players (say Alice) happens to be the first to query her physical system. This triggers the influence that brings information about x to Bob's system, propagating at speed v in the preferred frame. But now, even if $v > c$, such a signaling model cannot be tweaked to send information to the past, because the influence propagates at that speed in the preferred frame, not in each player's frame. Other signaling models may be built on more general definitions or more inventive uses of preferred frames.[4] None opens the possibility of sending messages to the past. However, it was proved that the process that generates the outputs in signaling models cannot have a Lorentz-invariant description (Gisin, 2011).

In summary: On the one hand, the greatest reason for fearing faster-than-light influences is not a concern; on the other hand, signaling models violate the principle

---

Toner-Bacon model simulating the singlet. Alice chooses her measurement in the $\hat{n}$ direction: If the recipe returns the output -1, with the Toner-Bacon bit of communication she can steer Bob's statistics to those of the desired state $|+\hat{n}\rangle$. If the output is $+1$, Bob's statistics will be steered to those of the state $|-\hat{n}\rangle$: Then Alice has to tell Bob to flip his output, and this is the role of the second bit of communication.

[3] A few steps later in the development of the theory, one also finds an energetic reason why *massive* objects cannot travel faster than light.

[4] I entered the field of nonlocality by helping Antoine Suarez to elaborate such a model: Each system sends out its superluminal signal in the rest frame of the measurement device it meets (Suarez and Scarani, 1997). The most clear prediction of this model was soon falsified in an experiment (Stefanov *et al.*, 2002) and it's while working on this that was born the idea discussed in the subsection 11.3.2.

of relativity. Although this principle is not a necessary pre-condition for rational enquiry, it has proved very fruitful, and applicable to the Universe as we know it. Abandoning it adds another item to the feeling of conspiracy: If there are preferred frames, it is not clear why they seem to be used only as "ether for Bell nonlocality."

All of this may start sounding wild, so let's refocus by recalling why people may be interested by such models: To save determinism. The simplest way to bring determinism back into physics would have been through local variable models. These were falsified by merely looking at the observable correlations. *May one falsify signaling models too on the sole observation of correlations? The answer is as close to a "yes" as it can be.* This is the object of the next subsection.

11.3.2   The need for infinite speed

For definiteness, we assume a signaling model using a single preferred frame, although the same argument can be exported to other models (Scarani *et al.*, 2014). If the speed v of the influence is infinite, any correlation can be distributed, because the information becomes instantaneously available everywhere. So, a model with infinite speed cannot be falsified by observation.

A *finite* speed defines an influence-cone in the preferred frame with the corresponding notion of space-like separation (of course, meaningful only in the preferred frame). Consider first a bipartite Bell scenario: If the players are sufficiently far apart, a player's output may be produced before the influence carrying information about the other player's input arrives at that location. According to the model, Bell nonlocality should disappear. It is refreshing that an alternative model provides testable predictions, and indeed experiments were designed to test it (Salart *et al.*, 2008): Observations showed that nonlocality was not dimmed—had it been, I would be writing very differently. But such experiments can only lead to a lower bound on v, and even that, only given a definition of the relevant events (which may be challenged, as discussed in subsection 1.5.4). A much stronger case against any signaling model with finite speed can be made by moving to *multipartite Bell* scenarios (Scarani and Gisin, 2002).

We consider N rounds of a Bell test in a three-partite Bell scenario. For the sake of the argument, we group all the inputs and outputs processes of each player as a single event denoted $\mathcal{E}_{A,B,C}$. A notation like $\mathcal{E}_A \overset{v}{\to} \mathcal{E}_B$ means that the influence carrying information about $\mathcal{E}_A$ can reach Bob's location prior to $\mathcal{E}_B$ taking place.

Besides the existence of the influence in a preferred frame, the signaling model is defined by two requirements:

(R1)   The observed behavior may deviate from an *expected behavior* $\mathcal{P}^*$ only when some of the events are space-like separated with respect to the influence-cone. For definiteness, we take the expected behavior to be the quantum one that the experimentalists may reconstruct from tomography; but the argument is independent of this specific choice.

(R2) The influence must remain *hidden*: In no configuration of events faster-than-light communication between players should be enabled.[5]

Now consider the following configurations of three measurements at different locations, with C located between A and B (Figure 11.1):

(I_B) $\mathcal{E}_A \xrightarrow{v} \mathcal{E}_B \xrightarrow{v} \mathcal{E}_C$.

(I_A) $\mathcal{E}_B \xrightarrow{v} \mathcal{E}_A \xrightarrow{v} \mathcal{E}_C$.

(*II*) $\mathcal{E}_A \xrightarrow{v} \mathcal{E}_C$ and $\mathcal{E}_B \xrightarrow{v} \mathcal{E}_C$, but $\mathcal{E}_A$ and $\mathcal{E}_B$ are outside each other's influence-cone; moreover, $\mathcal{E}_C$ is space-like separated from both $\mathcal{E}_A$ and $\mathcal{E}_B$ according to the usual light cones.

By definition, in configurations I_A and I_B we observe the expected behavior $\mathcal{P}^*$. We denote by $\mathcal{P}^{II}$ the behavior observed in configuration *II*.

Here comes the key observation. On the one hand, by timing his event $\mathcal{E}_B$, Bob can change the configuration between *II* and I_B. On the other hand, with signals traveling at the speed of light, information about $\mathcal{E}_C$ can reach Alice before information about $\mathcal{E}_B$. If

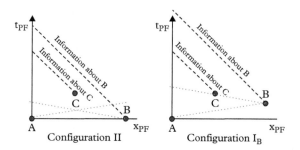

Figure 11.1 *The configurations of events used in the argument against hidden influences with finite speed. The space-time diagram is drawn in the preferred frame (PF) in which the influence is defined. For simplicity, the players' actions are depicted as instantaneous, so event A corresponds to Alice receiving x and producing a; and similarly for B and C. The grey dotted lines represent the hidden-influence cones, the black dashed lines the usual light cones. The argument goes as follows: On the one hand, Bob can switch between the two configurations by just delaying his event; on the other hand, in both configurations, information about B arrives to Alice later than information about C. If we assume that the players should not see any faster-than-light communication (i.e., the influences remain "hidden"), then the correlations A − C cannot depend on the configuration. A symmetric argument can be made for the correlations B − C. A contradiction is then found if those pairs of correlations enforce A − B to be nonlocal, since in configuration II the behavior of A − B must be local by definition of the model.*

---

[5] Recall that such influence-enabled faster-than-light communication would *not* allow sending information to one's past. So the cogency of this requirement is not as strong as usual. It can be seen as a desire to keep a "peaceful coexistence" with special relativity.

$\mathcal{P}_{AC}^* \neq \mathcal{P}_{AC}^{II}$, Alice would get to know which timing Bob has chosen *before* any information traveling at the speed of light could have carried this piece of information to her. This violates requirement R1. Thus, we need to enforce

$$\mathcal{P}_{AC}^{II} = \mathcal{P}_{AC}^*. \tag{11.1}$$

By the same argument, Alice can change the configuration between II and I_A by timing her event $\mathcal{E}_A$, so we need to enforce

$$\mathcal{P}_{BC}^{II} = \mathcal{P}_{BC}^*. \tag{11.2}$$

There is also a constraint on the third two-player marginal of $\mathcal{P}_{ABC}^{II}$: Since no influence connects the two events,

$$\mathcal{P}_{AB}^{II} \text{ must be local.} \tag{11.3}$$

The question is now: Can one find a Bell scenario and an expected behavior $\mathcal{P}^*$, such that the three constraints (11.1)–(11.3) on $\mathcal{P}^{II}$ are incompatible? In other words, we are asking if one can define marginal behaviors $\mathcal{P}_{AC}^*$ and $\mathcal{P}_{BC}^*$, such that *all* compatible $\mathcal{P}_{ABC}^*$ exhibit nonlocality in $\mathcal{P}_{AB}^*$ — in itself, an interesting problem in the mathematics of nonlocality.

For the configuration of events that we presented, examples of such $\mathcal{P}^*$ were indeed found, though they are not quantum behaviors (Coretti *et al.*, 2011). Quantum examples were found shortly later, first for a four-partite Bell scenario (Bancal *et al.*, 2012), then for a different configuration of events of a three-partite Bell scenario (Barnea *et al.*, 2013). There would be little added value in reporting the details here, those references are self-contained. If the reader wants to get a simple grasp of the problem, examples are very easy to find if one adds to (11.3) the requirement that $\mathcal{P}^{II}$ be quantum too (Scarani and Gisin, 2005).

In summary, *all models with a finite-speed influence that remains hidden* can be ruled out based on observation alone, without having to specify any detail (for instance, which is the preferred frame in which the influence propagates). An experimental validation would consist of a loophole-free Bell test exhibiting one of the $\mathcal{P}^*$. Notice that there is no need to argue how the event configuration looks like in the preferred frame:[6] One just has to be convinced that the behavior has not been realized by cheating via one of the loopholes. No such experiment has been reported, but the quantum behaviors $\mathcal{P}^*$ found in (Bancal *et al.*, 2012; Barnea *et al.*, 2013) can be realized with few

---

[6] This is exactly the strength of proper hidden-influence inequalities. In this respect, let us recall the work of (Jones *et al.*, 2005) described in subsection 5.3.3. There, it was proved that the Svetlichny inequality is satisfied if Alice can signal to both Bob and Charlie but these can't reply or communicate between them. This is exactly what a hidden-influence model would predict for the configuration of events studied in (Barnea *et al.*, 2013). But the Svetlichny inequality requires three-partite correlations: Its violation could falsify the hidden influence model only if one were sure of the configuration of events in the preferred frame.

projective measurements on four or three qubits, so nobody doubts that they could be observed.

11.4 Relaxing Measurement Independence

After relaxing no-signaling, we deal now with the relaxation of measurement independence. Measurement dependence (MD) means that, in each round, the process λ used by the players is correlated with the verifier's inputs (x, y). As usual, correlation does not say anything about causation, so various narratives are possible: The verifier gets to know the process and adapts the input, the players get to know the inputs and adapt their process, an external cause is influencing both the inputs and the process (this one will be used for randomness amplification in the next section).... The different narratives don't make any difference in the mathematics that follow.

11.4.1 Behaviors under measurement dependence

For a start, we need to change the definition of the observed behavior (2.5) that we used throughout the book, to take into account that the distribution of the processes λ now depends on the inputs:[7]

$$P(a, b|x, y) \overset{MD}{=} \int d\lambda P(\lambda|x, y) P(a, b|x, y, \lambda). \tag{11.4}$$

The behaviors that the players can produce with LV resources, i.e., those of the form

$$P(a, b|x, y) \overset{MDL}{=} \int d\lambda P(\lambda|x, y) P(a|x, \lambda) P(b|y, \lambda), \tag{11.5}$$

are called *measurement-dependent local (MDL)* behaviors. For a MDL behavior, $P(a|x, y)$ may depend on y, or $P(b|x, y)$ on x: Even if it's a convex mixture of local behaviors, the average behavior may violate the no-signaling condition. This is a consequence of MD and does not mean that the players can actually signal to each other.

The definition (11.4) of a behavior assumes i.i.d just like (2.5). In the case of measurement independence, we had eventually proved that the local polytope for non-i.i.d. behaviors is the same as that defined by i.i.d. behaviors (subsection 2.6.1). The analog result has not been proved for MD, and for some cases of constrained MD it has been proved not to hold (see the end of subsection 11.4.3).

---

[7] We also denote that distribution with P instead of Q and change the notation from $P_\lambda(a, b|x, y)$ to $P(a, b|x, y, \lambda)$. These cosmetic modifications, more suitable for the manipulations in this section, do not change the meaning of those objects. The only real mathematical difference is the correlation between λ and x, y.

11.4.2 Imposing restrictions on MD

Next, we must *impose restrictions* on MD. These could conceivably be motivated on physical grounds, but actual examples have not been found.[8] At the moment of writing, then, any *a priori* choice of restriction is arbitrary; and we don't get a real-life feeling for *a posteriori* results of the type "the observed behavior certifies Bell nonlocality even if this form of MD is allowed." Hoping that this unpleasant state be temporary, let's carry on.

All the restrictions studied so far are based on *quantifying MD according to some figure of merit*, then declaring that one tolerates at most this or that amount of MD. As figure of merit, in the early studies one finds the maximal deviation $M = \sup_{x,y,x',y'} \int d\lambda |P(\lambda|x,y) - P(\lambda|x',y')|$ (Hall, 2011) and the mutual information $H((\mathcal{X},\mathcal{Y}):\Lambda) = H(\mathcal{X},\mathcal{Y}) + H(\Lambda) - H((\mathcal{X},\mathcal{Y}),\Lambda)$ where $H(A) = -\sum_a P(a)\log P(a)$ and Λ is the alphabet of the process indices λ (Barrett and Gisin, 2011).

Subsequent studies have found it more convenient to quantify MD by bounding the $P(x,y|\lambda)$ according to

$$\ell \leq P(x,y|\lambda) \leq h \text{ for all } x,y,\lambda. \tag{11.6}$$

From $P(x,y) = \int d\lambda P(x,y|\lambda)P(\lambda)$, there follow the bounds

$$\ell \leq \min_{x,y} P(x,y), \quad h \geq \max_{x,y} P(x,y) \tag{11.7}$$

that reduce to $\ell \leq \frac{1}{M_A M_B} \leq h$ if $P(x,y) = \frac{1}{M_A M_B}$ for all (x,y), as frequently assumed.

Let us get a quick hold on these constraints. Consider first the $(2,2;2,2)$ Bell scenario. Setting $h = \frac{1}{3}$ and $\ell = 0$ opens the possibility that, in each round, the verifier promises not to query one of the four possible pairs of inputs, and the players know which one. In this case, the PR-behavior (9.3) can be realized as MDL behavior (Exercise 11.1). This implies that the whole no-signaling polytope can be realized with MDL behaviors, and Bell nonlocality in that scenario cannot be demonstrated. For a generic $(M_A, m_A; M_B, m_B)$ Bell scenario, we can be sure that the whole no-signaling polytope can be realized with MDL as soon as $h \geq \frac{1}{M_A + M_B - 1}$ if $\ell = 0$. Indeed, with these values, in each round the verifier may choose his inputs only among the $(M_A + M_B - 1)$ pairs $(x,y) \in \{(x',\bar{y}),(\bar{x},y')|x' \in \mathcal{X}, y' \in \mathcal{Y}\}$, for a given $\bar{x} \in \mathcal{X}$ and $\bar{y} \in \mathcal{Y}$. It can be proved that any behavior involving only those inputs is local (Thinh *et al.*, 2013). In all cases, notice that MD makes it impossible to certifiy nonlocality well before reaching $h = 1$, which is where the process and the inputs could be fully correlated.

The constraint (11.6) is also open to generalizations beyond i.i.d. that have been widely studied in the mathematical literature on sources of randomness. A natural generalization is the so-called *Santha-Vazirani (SV)* constraint:

---

[8] Here could be a candidate: In a lab, all the devices are usually connected to the city's electric grid; so, fluctuations in electric power affect both the devices that prepare the state and those that perform the measurements. However, in all the examples I am aware of, the effects of power fluctuations do not translate in the kind of MD that would be useful in Bell tests.

$$P(x_k, y_k | \lambda, x_{k-1}, \ldots, x_1, y_{k-1}, \ldots, y_1) \geq \ell \text{ for all } x, y, \lambda \text{ and for all } k. \tag{11.8}$$

This describes a non-i.i.d. process λ for which all pairs (x, y) have non-zero probability, at any round and whatever the past. Obviously this implies that no pair (x, y) has probability one either:[9] So, this process has randomness.

The loosest constraint is the so-called *min-entropy* constraint. Assuming that λ is a process that correlates N rounds, it reads

$$P(\underline{x}, \underline{y} | \lambda) \leq h_N \text{ for all } \underline{x} = (x_1, \ldots, x_N), \underline{y} = (y_1, \ldots, y_N), \lambda. \tag{11.9}$$

Contrary to the SV constraint, the min-entropy constraint is such that some of the sequences may have zero probability. For comparison with the i.i.d. case, one may define an effective single-round h_{eff} by setting $h_N \equiv h_{\text{eff}}{}^N$.

11.4.3 The MDL polytope, and one inequality

We are now in a position to discuss the set of MDL behaviors given a constrained MD. We devote most of our attention to the i.i.d. case with constraints (11.6). In order to use these constraints, we need to re-express MD behaviors (11.4) in terms of $P(xy|\lambda)$. This can be done using Bayes' rule $P(\lambda|xy) = P(xy|\lambda)P(\lambda)/P(xy)$. In the resulting expression, the constraints affect both the numerator and the denominator, since $P(xy) = \int d\lambda P(xy|\lambda)P(\lambda)$. It would be simpler to re-define behaviors as describing the joint distribution of outputs and inputs, rather than the conditional distribution, that is

$$P(a, b, x, y) \stackrel{MDL}{=} \int d\lambda P(\lambda)P(x, y|\lambda)P(a, b|x, y, \lambda). \tag{11.10}$$

For this new definition of behavior, the constraints (11.6) are linear. Then it can be proved that the set of i.i.d. MDL behaviors thus constrained forms a *polytope* for all values of ℓ and h (Pütz *et al.*, 2014; Pütz and Gisin, 2016). The explicit form of the polytope has been studied only for the $(2, 2; 2, 2)$ scenario; the number of facets varies according to the ranges of ℓ and h. The CHSH inequality is not a facet (although of course it can be studied in the context of MD, see Exercise 11.2).

Let us focus on one of the facets, possibly the simplest to write down, as well as the most studied. It is defined by the inequality

$$I_{MDL} = \ell P(0, 0, 0, 0) - h \left[P(0, 1, 0, 1) + P(1, 0, 1, 0) + P(0, 0, 1, 1) \right] \leq 0. \tag{11.11}$$

No short proof is known that this defines a facet, but let's prove that the inequality indeed holds for MDL behaviors. First, we rewrite it using (11.10):

---

[9] SV sources of randomness are usually presented for bits; the upper bound is then clear: $\varepsilon \leq P(b_k|\lambda, b_{k-1}, \ldots, b_1) \leq 1 - \varepsilon$.

$$I_{MDL} = \int d\lambda P(\lambda)\,[\ell P(0,0|0,0,\lambda)P(0,0|\lambda) - h\,[P(0,1|0,1,\lambda)P(0,1|\lambda)$$

$$+ P(1,0|1,0,\lambda)P(1,0|\lambda) + P(0,0|1,1,\lambda)P(1,1|\lambda)]]\,.$$

Now we can use (11.6) to bound $P(0,0|\lambda) \leq h$ in the first line, and $P(x,y|\lambda) \geq \ell$ for the three terms in the second line: Then

$$I_{MDL} \leq \ell h \int d\lambda P(\lambda)\,[P(0,0|0,0,\lambda) - P(0,1|0,1,\lambda) - P(1,0|1,0,\lambda) - P(0,0|1,1\lambda)]\,.$$

In the right hand side, there appears a normal Bell expression, without MD: In fact, it's the Eberhard form (2.33) of CHSH. If the initial $p(abxy)$ is MDL, that is $P(a,b|x,y,\lambda) = P(a|x,\lambda)P(b|y,\lambda)$, CHSH cannot be violated and $I_{MDL} \leq 0$ holds as claimed.

Let us now study the violation of I_{MDL} in quantum theory. By non-negativity of probabilities, no violation is possible if $\ell = 0$. As soon as $\ell > 0$, we know from Hardy's test (subsection 4.4.1) that I_{MDL} can be violated with quantum behaviors: Indeed, one can set $P(0,1,0,1) = P(1,0,1,0) = P(0,0,1,1) = 0$ while having $P(0,0,0,0) > 0$. This result holds independently of h. By noticing that $\ell = 0$ implies the possibility of excluding one of the pairs of inputs, these observations can be cast in a very appealing narrative: Some quantum behaviors can be proved to be nonlocal as long as all the inputs are possible in each round, however small the probability of some of them. It was noticed later that, in larger Bell scenarios with $M_A = M_B = M > 2$, some MDL inequalities are violated even for $\ell = 0$, provided $h \leq \frac{1}{M^2 - M + 1}$, which implies that at least 2 pairs of inputs must be possible in each round (Pütz and Gisin, 2016).

The systematic study of non-i.i.d. strategies has been attempted only for the (2, 2; 2, 2) scenario under the min-entropy constraint (11.9); even then, rather partial results were obtained (Tan *et al.*, 2016). One of them is worth mentioning because it shows, as claimed previously, that the MDL polytope changes significantly. Consider a $N = 2$ min-entropy constraint for the MD process. One can run this process and extract a single-round effective behavior $P(a, b|x, y)$ from the frequencies—in narrative terms, we can say that the verifier treats the process as being i.i.d. while it is not. For $h_{\text{eff}} \gtrsim 0.255$, a MDL strategy can give an effective behavior that violates I_{MDL}; whereas in the i.i.d. case, $h \approx 0.255$ forces a strictly non-negative $\ell = 1 - 3h \approx 0.235$. In summary, a criterion that demonstrates nonlocality for a i.i.d. MD scenario becomes inconclusive for a min-entropy constraint.

11.5   Randomness Amplification

11.5.1   DI certification and measurement dependence

The possibility of DI certification follows from the certification of nonlocality. The latter is usually done under strict measurement independence and no-signaling (as we assumed in Part II and section 9.3). But the certification of randomness has also been the object of

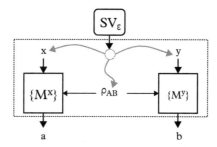

Figure 11.2 *The idea of randomness amplification. The output of a Santha-Vazirani source (11.13) is used to choose both the measurement settings and the state to be used in a Bell test, thus introducing clear measurement dependence. The goal is to prove that, even in this situation, the outputs of the Bell test can define a process that is arbitrarily close to a perfect coin (at least for some values of ε). Thus, the Bell test (dashed box) acts as an unseeded extractor of a SV source, something that is impossible in information theory with classical resources.*

exploratory studies when nonlocality is certified under partial measurement dependence (Koh *et al.*, 2012) or bounded signaling (Silman *et al.*, 2013).

In this section, we rather give a different twist to the relation between nonlocality and randomness in the presence of measurement dependence. Discussing randomness extraction (subsection 8.2.2) we pointed out that seeded extraction is possible for fairly generic sources, whereas deterministic extraction is only possible in limited cases. In particular, the randomness of SV sources (11.8) cannot be extracted deterministically. Colbeck and Renner (2012) proved that the randomness of an SV source *can be extracted* by using the original source as input of a Bell test and taking the outputs as the new source (Figure 11.2). This result could be presented as another example in which quantum information processing achieves something that is impossible in classical information processing. I prefer to read it as a proof that a Bell test generates genuine independent randomness, as needed to extract randomness from a SV source.

11.5.2 Setting the stage

The proof of Colbeck and Renner uses the chained inequalities that we studied in subsection 4.2.3. The inequality uses $2M$ pairs of settings, so we can assume that the source of randomness picks only those pairs. On average over many rounds, each pair is chosen with the uniform distribution

$$P(x, y) \equiv Q(x, y) = \frac{1}{2M}. \tag{11.12}$$

Claire however knows that the settings are chosen with a SV source defined by

$$\frac{1 - \varepsilon}{2M} \leq P(x, y | \Lambda) \leq \frac{1 + \varepsilon}{2M} \tag{11.13}$$

where Λ is a shorthand notation for both the process λ chosen in that round (this is measurement dependence) and the inputs of the previous rounds.

In chapter 8, we considered randomness for someone that has the most precise description of the process in each round allowed by quantum theory; and we focused mostly on secret randomness, i.e., for an adversary outside the secure location. Here, the goal is different: We start with a source of randomness, and secrecy is not a direct concern. In this situation, von Neumann extraction would be possible if the source were i.i.d. (subsection 8.2.2). We are going to show how using nonlocality one can obtain a i.i.d. source of bits starting with the SV source (11.13). This step is called *randomness amplification*; extraction would then be completed by von Neumann's or other procedures (although in this particular example the output bits are already unbiased).

The figure of merit is the variational distance (9.12) between the actual distribution and the desired one. Let us set the distance between the SV source (11.13) and the unbiased i.i.d. source (11.12) to[10]

$$\mathcal{D}[P_{\mathcal{X},\mathcal{Y}|\Lambda}, Q_{\mathcal{X},\mathcal{Y}}] \le \frac{\varepsilon}{2}. \tag{11.14}$$

11.5.3   Derivation

The proof of the possibility of randomness amplification uses similar tools as in subsection 9.4.2: A bound on a variational distance through the chained inequality. Again we refer to Appendix G.7 for the manipulations of the variational distance.

We consider the chained inequality in the form (4.7) for given W:

$$C'_{M|W} = P(a = b|1, M, \Lambda) + \sum_{x,y}{}' P(a \ne b|x, y, \Lambda)$$

where $a, b \in \{0, 1\}$ and where the prime on the sum selects only the pairs of settings that enter the inequality (those that the SV source can select). We can follow the proof of subsection 9.4.2 till (9.14), which reads here $C'_{M|W} \ge \mathcal{D}[\tilde{P}_{A|x=1,\Lambda}, P_{A|x=1,\Lambda}]$ that is

$$C'_{M|\Lambda} \ge 2\mathcal{D}[P_{A|x=1,\Lambda}, Q_A] \tag{11.15}$$

where $Q_A(a) = \frac{1}{2}$ represents the i.i.d. source with unbiased marginals.

Now we have to relate the bound (11.15) on $C'_{M|\Lambda}$ to the observed value of C'_M. This is not immediate because of measurement dependence: Indeed, $C'_{M,\mathrm{obs}} \ne \sum_\Lambda P(\Lambda) C'_{M|\Lambda}$, but rather

---

[10] Notice that (Colbeck and Renner, 2012) defined $\mathcal{D}[P_{\mathcal{X},\mathcal{Y}|\Lambda}, Q_{\mathcal{X},\mathcal{Y}}] \le \varepsilon$, so there is a factor of two between our and their definition.

$$C'_{M,\text{obs}} = \sum_{\Lambda} P(\Lambda|1,M)P(a=b|1,M,\Lambda) + \sum_{x,y}{}' P(\Lambda|x,y)P(a \neq b|x,y,\Lambda). \qquad (11.16)$$

By manipulating this expression:

$$C'_{M,\text{obs}} = \sum_{\Lambda} P(\Lambda|1,M)\left\{ P(a=b|1,M,\Lambda) + \sum_{x,y}{}' \frac{P(\Lambda|x,y)}{P(\Lambda|1,M)}P(a \neq b|x,y,\Lambda)\right\}$$

$$\geq q_M(1,M)\sum_{\Lambda} P(\Lambda|1,M)C'_{M|\Lambda}$$

$$\geq 2q_M(1,M)\mathcal{D}[P_{\mathcal{A},\Lambda|x=1,M}, \mathcal{Q}_{\mathcal{A}} \times P_{\Lambda|x=1,M}] \qquad (11.17)$$

where for the last line we have used (11.15) together with the no-signaling condition. We have defined

$$q_M(1,M) = \min_{x,y,\Lambda} \frac{P(\Lambda|x,y)}{P(\Lambda|1,M)} \geq [(1-\varepsilon)/(1+\varepsilon)]^{2\log_2 M} \qquad (11.18)$$

with the minimum taken on the pairs of settings with non-zero probability, and where the bound follows from (11.14).

Finally, let us insert into (11.17) the best quantum value (in this case, a mimimum) $C'_{M,\text{obs}} = 2M\sin^2\frac{\pi}{4M} \lesssim \frac{\pi^2}{8M}$:

$$\mathcal{D}[P_{\mathcal{A},\Lambda|x=1,M}, \mathcal{Q}_{\mathcal{A}} \times P_{\Lambda|x=1,M}] \leq \frac{\pi^2}{16}\left(\frac{1+\varepsilon}{\sqrt{2}(1-\varepsilon)}\right)^{2\log_2 M}. \qquad (11.19)$$

The r.h.s. tends to zero for large M for $\varepsilon \leq (\sqrt{2}-1)^2 \approx 0.172$. In conclusion, if the initial bias is not too large, the process that produces Alice's outputs for $x=1$ is as close as one wishes to an unbiased i.i.d. coin.

One of the crucial elements of the proof is the fact that the quantum bound and the no-signaling bound of the chained inequality coincide. We know other inequalities with that property, e.g., the MABK inequalities for an odd number of players (section 5.2). With this insight, it did not take long after the Colbeck-Renner breakthrough to find a five-player Bell test that would allow randomness amplification for any $\varepsilon < \frac{1}{2}$ (Gallego *et al.*, 2013). Both results are very sensitive to small deviations from the perfect quantum value. A series of further improvements followed, reviewed in (Pivoluska and Plesch, 2014). At the moment of writing, possibly the best result is the randomness amplification of a SV source based on the inequality I_{MDL} (11.11) obtained using the entropy accumulation theorem (Kessler and Arnon-Friedman, 2017).

....

EXERCISES

Exercise 11.1. *Assume that $h = \frac{1}{3}$ and $\ell = 0$ in (11.6). For this case of measurement dependence, design a LV model that reproduces the PR-behavior (9.3).*

Exercise 11.2. *We study CHSH in the presence of MD. We have said that CHSH is not a facet of the MDL polytope, so it is not the most robust inequality against MD. But it has been the most tested one, so it is interesting to ask to which extent the reported observations are tolerant against MD.*

1. *Prove that the observed correlation coefficients E_{xy} are given by*

$$E_{xy} = \frac{1}{P(x,y)} \int d\lambda P(\lambda) P(x,y|\lambda) E_{xy}^{\lambda} \tag{11.20}$$

   where $p(x,y) = \int d\lambda P(\lambda) P(x,y|\lambda)$. Hint: Out of N rounds, each λ is used in $N_\lambda = p(\lambda)N$ rounds on average; and E_{xy}^λ is probed in $P(x,y|\lambda)N_\lambda$ rounds on average.

2. *Consider the following MD strategy: There are four λ's denoted by $\lambda_{x'y'}$, $x',y' \in \{0,1\}$, each used with probability $P(\lambda_{x'y'}) = \frac{1}{4}$. They are defined by*

$$P(x,y|\lambda_{x'y'}) = p + (1-4p)\delta_{x,x'}\delta_{y,y'}$$

   for $0 \leq p \leq \frac{1}{3}$. Find that in this case the CHSH expression is given by

$$S = \sum_{x'y'} \sum_{x,y} (-1)^{xy} P(x,y|\lambda_{x'y'}) E_{xy}^{\lambda_{x'y'}}.$$

3. *Convince yourself that the local bound of this inequality has already been studied in Exercise 1.4, with $q = 1 - 3p$ [refer to (Koh et al., 2012) for the study of its violation in quantum theory: In particular, as long as it is not achievable with LVs, its maximal violation self-tests the maximally entangled state of two qubits].*

12

Epilogue

Seul l'avenir nous dira de quoi le futur sera fait.
Only future will tell what tomorrow has in stock.

Attributed to C. Constantin

Many popular accounts on nonlocality build on the catchy slogan "Einstein was wrong." He may well have been; but prescient, he surely was: He realized that, if local hidden variables are not the explanation, our view of physics or of nature will be deeply affected. After Bell, we can only choose *how* it will be affected.

We can adopt the "orthodox" view, that decrees that one should not try to describe individual events. Individual events are used to update one's knowledge; when a usable pattern emerges, in the form of a quantum state, it enables future predictions, all of statistical nature. This is "just standard quantum mechanics," as some of my colleagues like to say. This should not make it less shocking: individual events have no place in physics, the description of a round of an experiment is unspeakable.

If we want physics to say something about those individual events, nonlocality kicks in fully, with influences that propagate at infinite speed in a preferred frame. Granted, we may change the wording (infinitely rigid ether, quantum potentials, pilot waves, preferred foliation . . .) or the narrative (add some convenient extra dimensions, massless wormholes, or other science fiction constructs). But the basic fact remains: A description of those individual events won't fit in the space-time setting we are familiar with.

Some take refuge in bigger conspiracies, from incredible measurement dependence to full blown superdeterminism. They claim that they can live meaningful lives, and even have moral duties like that of promoting science, while adopting such a worldview. I won't be able to follow them there, but this may just mean that I have been programmed differently.

One of my few certainties is that the future of physics won't be milder, that more shocks are in stock for the future. Nonlocality was discovered as one of the many phenomena predicted through quantum theory, but it has the potentiality to acquire the status of principle. Will it be the solid base from which to take the next leap?

Bell Nonlocality. Valerio Scarani. © Valerio Scarani 2019. Published in 2019 by Oxford University Press.
DOI: 10.1093/oso/9780198788416.001.0001

Appendix A
History Museum

This appendix covers a few of the milestones in the debate about indeterminism in quantum theory. The pre-history of the mathematical object "Bell inequalities" starts with George Boole's *The Laws of Thought* of 1854; it has been reviewed in (Pitowsky, 1989).

A.1 Heisenberg's Uncertainty Relations and the Question of Indeterminism

As we said in the main text, traditional textbooks argue in favour of intrinsic indeterminism by invoking Heisenberg's uncertainty relations: The better one can predict the result of a measurement of position, the worse one can predict the result of a measurement of momentum. The orthodox interpretation of quantum theory does consider the uncertainty relations as a manifestation of intrinsic indeterminism. However, the cogency of the argument is limited by several considerations:

- *Ambiguous interpretation.* The orthodox interpretation does not impose itself. When pressured to explain uncertainty to the general public, Heisenberg himself attributed it to an alleged intrinsic limitation in the measurement process. His explanation based on the "Heisenberg microscope" has been repeated countless times, also by authors who were otherwise endorsing the orthodox view on quantum indeterminism. Alternatively, one may also consider intrinsic fluctuations in the preparation: The source may produce a sharp pair (x, p) in each round, but there is something that prevents it from always producing the same pair.

- *Reliance on quantum theory.* Let us anyway assume the orthodox interpretation, that the values of some variables are really not sharply defined. Without quantum theory, we wouldn't know which form of unpredictability this gives rise to. The "natural" guess would be that every physical quantity has some intrinsic noise, but this is not what nature seems to have opted for. Rather, at least two physical quantities must be considered to see some intrinsic noise.[1] And it's impossible to quantify the amount of intrinsic noise *a priori*: At which point would one have to admit "I can't do better, I have reached the limit imposed by nature"?

- *Reliance on characterization of the devices.* Let us even assume the validity of quantum theory, so that we know what to test and which bounds to expect. The uncertainty relations cannot

---

[1] Gravity may be an exception as hinted by Matvei Bronstein in the 1930s with the following argument. Bohr and Rosenfeld had shown that a sharp measurement of field components is possible in principle: It uses a probe particle such that $\frac{e}{m} \to \infty$, where e is the charge associated to the field and m the mass. But for gravity, the charge associated to the field is the mass itself.

be tested without a very rigorous characterization of the devices: Which degree of freedom they measure, and how well calibrated they are. Nobody will doubt, for instance, that if two orientations of a Stern-Gerlach magnet are believed to be exactly orthogonal but are not, the uncertainty relations for orthogonal directions may be violated. At a similar level of triviality, if the apparatus that measures momentum is not switched on, its pointer will indicate $p = 0$ in each round, leading to estimate $\Delta p = 0$ irrespective of Δx. An unwitting combination of bad calibration and residual excessive noise may lead to a very tight "verification" of the uncertainty relations.

Because of all this, it is not surprising that the claim of intrinsic unpredictability was relegated to a question of interpretation prior to Bell nonlocality. In particular, it was certainly possible to consider the indeterministic formalism as an effective tool, to be replaced one day by a more comprehensive theory that would restore determinism.

A.2   The EPR Argument

The article by Einstein, Podolsky, and Rosen (Einstein, Podolsky, and Rosen, 1935) is often cited as "EPR paradox." I prefer to call it *EPR argument*, as I don't find anything paradoxical with it— unless one wants to call "paradox" the contradiction between pre-determined values and quantum theory.

EPR considered the state of two particles, each with its own position and momentum operators satisfying $[x_j, p_j] = i\hbar$. Since $[x_1 - x_2, p_1 + p_2] = 0$, one could try and define the state that satisfies both

$$x_1 - x_2 = d \ \text{ and } \ p_1 + p_2 = u \tag{A.1}$$

with $d, u \in \mathbb{R}$. As usual, the requirement of sharp position and momentum leads to unnormalizable wave-functions; but it is sufficient to add an arbitrary amount of spread to define a valid state.

EPR then reasoned as follows: If the measurement of the position of particle 1 yields $x_1 = x$, we know for sure that the result of a measurement of position of particle 2 would be $x_2 = x - d$. Similarly, if the measurement of momentum of particle 2 yields $p_2 = p$, we know that a measurement of momentum on particle 1 would have certainly given $p_1 = u - p$. This is perfectly valid quantum theory.[2] But then, they asked, how can such perfect correlations be accounted for by the orthodox interpretation, according to which the results of the measurements do not pre-exist? The whole book has been devoted to this question, so we won't dwell on it here.

It is interesting to notice that, even if EPR had been aware of Bell inequalities and had tried to violate them, they were handling one of the most difficult states for the task. Surely it's a pure entangled state, so a violation can be found for suitable measurements (Banaszek and Wódkiewicz, 1998). But the state has a positive Wigner function: So, as long as both particles undergo arbitrarily many measurements of the type $\cos \theta x + \sin \theta p$, there exist a LV model that describes all the statistics. In particular, the statistics of $(x_1, p_1; x_2, p_2)$ do have a LV description, as the EPR reasoning shows.

---

[2] In particular, contrary to a widespread misunderstanding in popular accounts, the uncertainty principle is *not* violated. Indeed, given the EPR state, the distribution of x_1 is uniformly spread over the line, so statistics would show a very large variance; and the same yields for x_2, p_1, and p_2. So $\Delta x_j \Delta p_j$ is arbitrarily large for both $j = 1, 2$.

A.3   Von Neumann's Observation on Pre-Established Values

Von Neumann's observation on pre-established values is part of his derivation of the Born rule (a derivation later superseded by Gleason's theorem). Once extracted, the argument is ultimately rather simple; in fact, it's used in several basic introductions to quantum theory. Let me phrase it with spins. On the one hand, a measurement of a spin $\frac{1}{2}$ will yield ± 1 (in some units) in any direction. On the other hand, a spin is a vector, so the component in the $\hat{d} = \frac{\hat{x}+\hat{y}}{\sqrt{2}}$ direction should satisfy $S_{\hat{d}} = \frac{S_{\hat{x}}+S_{\hat{y}}}{\sqrt{2}}$. But if we assign pre-established values $S_{\hat{x}} = +1$ or -1, and $S_{\hat{y}} = +1$ or -1, there is no chance that $S_{\hat{d}} = +1$ or -1 too.

This is an important observation on deterministic local hidden variable models: If there are pre-established values, their algebra can't respect the tensorial character of the physical objects that they describe. As it often happens, a complex statement was transmitted as a much simpler slogan, namely: Local hidden variables are ruled out. It was Bell (1966) that pointed out the flaw in the slogan: To measure the spin in the $\hat{d}$ direction, the Stern-Gerlach magnet must be oriented in a direction that is macroscopically distinguishable from any other. As such, it is a completely different measurement, so there is nothing irrational in assuming that its output is not algebraically related to that of other measurements.

We'll bring up similar topics in Appendix D dedicated to local variable models.

A.4   Bell's 1964 Inequality

The last piece of our museum is Bell's original inequality. Equation (15) in (Bell, 1964) reads

$$1 + P(\vec{b}, \vec{c}) \geq |P(\vec{a}, \vec{b}) - P(\vec{a}, \vec{c})| \tag{A.2}$$

where $P(\vec{a}, \vec{b})$ is a notation for the correlation coefficient $\langle a_x b_y \rangle$ with $\vec{a} \equiv \hat{a}_x$ and $\vec{b} \equiv \hat{b}_y$. We see that Alice uses measurement directions $(\vec{a}, \vec{b})$, while Bob uses measurement directions $(\vec{c}, \vec{b})$. In his derivation, Bell assumed that

$$P(\vec{b}, \vec{b}) = -1 \tag{A.3}$$

holds, as for the singlet state in quantum theory. Under this assumption, it is easy to check that the inequality (A.2) is a version of CHSH.

Because of this assumption, Bell's work already proved that the predictions of quantum theory are at odds with any local hidden variable theory. However, perfect anti-correlations are never observed in experiments: In order to perform a Bell test, one needs inequalities like CHSH that are independent of such idealizations.

Appendix B
Experimental Platforms: A Reading Guide

This appendix provides a quick overview of the most important experimental implementations of Bell tests. We don't go into any detail and just aim at lowering the potential barrier needed to read the research articles. The few cited experimental articles and the reviews (Genovese, 2005; Pan *et al.*, 2012) should be sufficient to start looking for more detailed information.

B.1 Photons

The vast majority of Bell tests have been conducted with *pairs of entangled photons*. The entanglement is often in polarization, although other degrees of freedom have been explored (subsection B.1.2). The main challenge in conducting a proper Bell test is the efficiency of single photon counters: Conventional semi-conductor detectors have efficiencies in the realm $10 - 50\%$, too low to close the detection loophole even if there were no losses. The loophole was eventually closed using superconducting detectors, which must be operated at low temperatures (in dilution refrigerators).

B.1.1 Sources

The early experiments used an *atomic cascade* as source (Aspect *et al.*, 1982b; Aspect *et al.*, 1982a). The mechanism is intuitively clear. The electronic state of an atom is prepared in an excited state that can decay through two possible paths: First emit a photon polarized left-circular then one polarized right-circular, or the opposite. If the first and the second transition have the same wavelength in both paths, the two processes are indistinguishable and the photons are entangled in polarization.

The big inconvenience of this source is that the emission can occur in a very large solid angle, so only rarely are the photons emitted in the directions that are collected by the measurement devices. This is why this type of source was completely abandoned in favor of sources based on *spontaneous parametric down-conversion*. In this process, light from a laser at frequency ω_L impinges from a direction $\hat{k}_L$ on a crystal that has non-linear optical response. The three-wave-mixing component $\chi^{(2)}$ can give rise to pairs of excitations that satisfy the phase-matching conditions $\omega_a + \omega_b = \omega$ and $\vec{k}_a + \vec{k}_b = \vec{k}_L$. By solving these equations given the dispersion relation in the crystal $\frac{\omega}{|\vec{k}(\omega)|} = \frac{c}{n(\omega)}$, the spatial modes a, b in which the photons are emitted are sharply defined for a given frequencies (ω_a, ω_b). It's good to keep in mind that the down-conversion efficiency is typically 10^{-4} (that is, the process produces one pair of photons per 10^4 photons in the pump beam).

Next we should discuss in which degrees of freedom can such photons be entangled.

B.1.2 Entangled degrees of freedom

Most experiments with photons, including two of the loophole-free ones (Giustina *et al.*, 2015; Shalm *et al.*, 2015), used entanglement in *polarization*. To explain its origin, one has to have a closer look into the possible non-linear crystals and their polarization properties.

- In *Type-II* down-conversion, there exists a basis (H, V) such that one of the down-converted photons has the same polarization of the pump and the other one opposite polarization. One can find suitable directions such that the phenomenological Hamiltonian reads

$$H_{II} = g\left(\alpha_H\, a_H^\dagger b_V^\dagger + \alpha_V\, a_V^\dagger b_H^\dagger\right) \tag{B.1}$$

  where $\alpha_{H,V}$ is the amplitude of the pump laser in that polarization mode and g is a coupling constant proportional to $\chi^{(2)}$ and roughly proportional to the length of the crystal. Thus, the first excitation of the output field is the two-photon state $\cos\theta\,|H\rangle_a|V\rangle_b + \sin\theta\,|V\rangle_a|H\rangle_b$ with $\cos\theta = \frac{\alpha_H}{\sqrt{\alpha_H^2 + \alpha_V^2}}$ (in this elementary presentation we can assume the amplitudes to be real). This was the type of entanglement used in (Weihs *et al.*, 1998).

- In *Type-I* down-conversion, there exist a polarization direction D such that pump light of that polarization may be converted into two photons with the orthogonal polarization $D^\perp$. The down-conversion process is captured phenomenologically by the Hamiltonian

$$H_I = g\alpha_D\, a_{D^\perp}^\dagger b_{D^\perp}^\dagger. \tag{B.2}$$

  Entanglement is produced by a serial arrangement of two crystals, the first aligned such that $D = H$, the second such that $D = V$. Then, for a pump polarized along $\cos\theta H + \sin\theta V$, the first excitation of the output field will be the two-photon state $\cos\theta\,|V\rangle_a|V\rangle_b + \sin\theta\,|H\rangle_a|H\rangle_b$.

As first noticed by Franson (1989), the down-conversion process itself defines another type of entanglement, that has been called *energy-time* entanglement or, in its discrete version, *time-bin* entanglement. This type of entanglement was used for instance in (Tittel *et al.*, 1998). The idea is the following: The coherence of the down-conversion process is defined by the coherence time of the pump τ_P. However, conservation of energy requires that the two down-converted photons are created within a very short time τ_c from each other. If $\tau_P/\tau_c \equiv N \gg 1$, which is usually the case, the output is multimode, so the phenomenological Hamiltonian (B.2) should rather read

$$H_I \approx g\alpha\sum_{j=1}^{N} a_j^\dagger b_j^\dagger \tag{B.3}$$

where we omitted the mention of the polarization and opted for a discrete choice of modes for notational simplicity. It follows that the first excitation of the output field is the two-photon state $\sum_{j=1}^{N} |j\rangle_a|j_b\rangle$, where j now refers to the time mode. This state is highly entangled, the only remaining challenge consists in reading that entanglement out. This can be done by unbalanced interferometers. Without describing the details, let me mention that the feasible implementations at the moment of writing cannot close the detection loophole, because the design of the interferometers has intrinsic losses.[1]

B.2 Atomic Degrees of Freedom

Trapped ions and atoms can be prepared in entangled states of their internal degrees of freedom (usually fine or hyperfine energy levels), after which one can observe violation of Bell inequalities. From the point of view of the loopholes, atomic systems are the opposite of photons: It's easy to close the detection loophole but hard to close the locality loophole.

The detection is based on driving a transition between one of the levels to be measured, say the ground state, and a third level. If the system was in the ground state, this process will generate easily detectable fluorescence. If the system was in the excited state, no light is emitted. Measurements in different bases can easily be effected by applying a Ramsey pulse before the detection (exactly the analog of rotating the polarization prior to measuring in a fixed basis). Because fluorescence detection is very efficient, the detection loophole was first closed with ions (Rowe *et al.*, 2001). A Bell violation with entangled ions in different chambers led to the first demonstration of randomness generation (Pironio *et al.*, 2010).

In order to close the locality loophole, techniques had to be devised to entangle two very distant atomic systems. The obvious way to do is to swap entanglement from a pair of entangled photons. While this proposal is easy to understand, its implementation is challenging. The main point is to *herald* the creation of the entanglement of the atomic systems, so that one performs the Bell test only when entanglement is known to have been created. Indeed, as we have seen, down-conversion is a probabilistic process; and even when entangled photons are produced, they may be lost in transmission or not couple to the atomic system. The first successful demonstration was the loophole-free Bell test of Delft, where the atomic degrees of freedom were electronic states of NV-centers in diamonds (Hensen *et al.*, 2015). Weinfurter's group later reported the same achievement with hyperfine states of neutral atoms (Rosenfeld *et al.*, 2017).

B.3 How to Address the Detection Loophole

As discussed in subsection 1.5.2, from the point of view of the verifier, it is trivial to close the detection loophole: He just needs to elicit an answer for each query. The challenge is in the design of the experimental setup. The detection loophole is also crucial in device-independent characterization, notably because it can be activated by sheer mistake—it has actually happened, and was duly reported (Tasca *et al.*, 2009).

This section describes a toy model that elucidates how one addresses the detection loophole; a much more detailed modeling will be required for any specific setup.

B.3.1 Setting up the notions

Simple theoretical estimates address the detection loophole by studying a single figure of merit, usually called *detection efficiency* and denoted η. This figure of merit actually captures what is called

---

[1] At some point, this obvious observation has given rise to a uselessly acrimonious discussion. I prefer to consign it, together with the corresponding references, to oblivion.

system efficiency in the experimental jargon: It's not the efficiency of the detector alone, but it includes the probability that the signal from the source is not lost before reaching the detector. To keep the number of parameters minimal, here we assume that η is the same for all the detectors of both players.

For definiteness, we consider a measurement setup with two conclusive outputs ("detections") and a third non-conclusive output ("no-detection"). The two conclusive outputs are naturally going to be associated with two outputs of the Bell test. But the players must give outputs in the Bell test even in the event of no-detection. There are two possible options:

- "No-detection" is treated as a third output. This leads to a Bell scenario with $m_A = m_B = 3$.
- "No-detection" is associated to one of the two outputs according to a pre-established recipe (often a deterministic assignment[2]). This leads to a Bell scenario with $m_A = m_B = 2$.

In all the examples that have been reported, these two options lead to similar, when not identical, thresholds for closing the detection loophole.

From now on, we focus on $M_A = M_B = 2$. The thresholds are comparable for other Bell scenarios with small number of inputs and outputs; for significantly reduced thresholds, it seems that one would have to go to unrealistic implementations. I refer to subsection VII.B.1.c of (Brunner *et al.*, 2014) for detailed references, and to (Pál and Vértesi, 2015) for further technical results. This is why all the experiments that closed the detection loophole performed to date have used the CHSH inequality.

B.3.2 Efficiency for the (2, 2; 2, 2) scenario

Suppose that Alice and Bob share a perfect maximally entangled state of two qubits and perform the measurements that saturate the Tsirelson bound. When both particles are detected, which happens with probability η^2, their data will show $S = 2\sqrt{2}$. When only one of the particles is detected and not the other, the data will show no correlations, because one of the outputs is the result of a measurement on half of a maximally entangled state, which is correlated only to its twin. So, a fraction $2\eta(1 - \eta)$ of the data will contribute with $S = 0$. Finally, when neither particle is detected, the most one can hope for is the local bound $S = 2$. This is easy to achieve: The players may just decide to output deterministically $+1$ in case of no-detection. Thus, the total statistics will give

$$S(\Psi^-, \eta) = \eta^2 2\sqrt{2} + (1 - \eta)^2 2 \tag{B.4}$$

and the condition $S(\Psi^-, \eta) > 2$ for violation translates into the threshold efficiency

$$\eta > \eta_{th}(\Psi^-) = \frac{2}{\sqrt{2}+1} \approx 82.8\%. \tag{B.5}$$

For good measure let us redo the estimate of the threshold efficiency with the CH form (2.30) of the CHSH inequality (as mentioned in subsection 2.5.2, this was the original calculation). Since this inequality only involves the outputs $a = b = 0$, one can group the no-detection events with

---

[2] In this case, in setups where each output is associated to a detector (as is the case for photons), one can have only one detector per player: Whenever it does not click, the other output is recorded.

the events $a = 1$ on Alice and $b = 1$ on Bob; but the number of such events must be counted in the denominator when computing frequencies. Then one would observe $P(0,0|x,y) = \eta^2 \frac{1}{4}(1 + (-1)^{xy}/\sqrt{2})$ and $P(a = 0|x = 0) = P(b = 0|y = 0) = \eta/2$, leading to $S_{CH} = \eta^2 \frac{\sqrt{2}+1}{2} - \eta$. As expected, the threshold efficiency for violation $S_{CH} > 0$ is the same (B.5).

B.3.3   A surprising optimization

As we have seen, assuming that the players share a maximally entangled state leads to the maximal value of S by the two-detection events, but implies a contribution $S = 0$ by the one-detection events. Surprisingly, Eberhard (1993) found that it is beneficial to trade a decrease in the first for an increase in the second. We present this result using the usual form of CHSH, although of course in the original paper it was found using the Eberhard form (2.33).

Consider non-maximally entangled pure two-qubit states (3.15) and leave the measurement directions free to start with. In a similar way as previously mentioned, the value of the CHSH parameter takes the form

$$S(\Psi(\theta), \eta) = \eta^2 S_2 + 2\eta(1 - \eta)S_1 + (1 - \eta)^2 2 \tag{B.6}$$

where the values of the S_j depend on the measurement directions. The threshold efficiency for every θ is then given by

$$\eta_{th}(\theta) = \min_{\hat{a}_0, \hat{a}_1, \hat{b}_0, \hat{b}_1} \frac{4 - 2S_1}{S_2 - 2S_1 + 2}. \tag{B.7}$$

The numerical optimization shows that $\eta_{th}(\theta)$ decreases monotonically with θ, from (B.5) for $\theta = \pi/4$ down to $\eta_{th}(\theta \to 0) = 2/3$. Besides, for $\theta < \pi/4$, the measurement settings that minimize the threshold efficiency are not those that give the maximal value (3.17) of S_2. Thus, we reach the conclusion that the detection loophole for CHSH is best closed with *weakly entangled* states,[3] and for measurements that *do not maximize* the contribution of the two-detection events.

The two loophole-free Bell tests conducted with photons used the Eberhard trick (Giustina et al., 2015; Shalm et al., 2015), whereas in the one conducted with NV-centers the efficiencies were high enough to detect the violation by a maximally entangled state (Hensen et al., 2015).

---

[3] It must be stressed that we are speaking of *pure* states, not of any weakly entangled state. In practice, if θ is too small, a low amount of noise will be enough to prevent any violation.

Appendix C
Notions of Quantum Theory Used in this Book

Undergraduate quantum physics is taken for granted, including the notation, the usage and the meaning of those mathematical objects (e.g., bras and kets, Hermitian operators, unitary operators, projectors …). This appendix lists the non-elementary notions of quantum kinematics that are used in the main text. For more comprehensive definitions and all the proofs, readers can refer to[1] (Nielsen and Chuang 2000; Schumacher and Westmoreland 2010; Watrous 2018).

C.1 States and Measurements

C.1.1 Definition of quantum theory

The content of this subsection is among the basic material that is supposed to be known. It is sketched here because it is directly related to some discussions in the main text, notably chapter 10.

Quantum theory is defined by the fact that physical systems are described by *vector spaces*. Barring foundational discussions, these vector spaces are complex. The finite-dimensional ones are all isomorphic and denoted $\mathbb{C}^d$ where d is the dimension. Only one infinite-dimensional vector space is used in quantum theory, the space of square-integrable functions $L^2(\mathbb{R})$. Practitioners refer to all these vector spaces as *Hilbert spaces*, borrowing the name of the infinite-dimensional case.[2] The state of a composite system is described by the *tensor product* of the Hilbert spaces of the subsystems.

Physical properties are represented by *subspaces* of the vector space, with the rule that distinguishable physical properties are associated to orthogonal subspaces. A *pure state* is a state of maximal knowledge, so it corresponds to the smallest non-trivial subspace: A one-dimensional ray. This ray can be unambiguously represented by the corresponding projector. Alternatively, one can represent it with a vector $|\psi\rangle$ lying on it, keeping in mind that any vector that differs by a constant represents the same state. A state ρ that does not describe maximal knowledge is called *mixed state* because it can always be represented as a convex sum of projectors on pure states.

---

[1] At the time of writing, Watrous' book is available online: https://cs.uwaterloo.ca/ watrous/TQI/. Another classic resource, to date unpublished, are Preskill's lecture notes available at http://www.theory.caltech.edu/people/preskill/ph229/.

[2] Mathematicians did not find it necessary to give a name to such a trivial object as a finite-dimensional vector space. Also, to be precise, the infinite-dimensional vector space is a *rigged* Hilbert space.

By contrast, in a *classical theory*, physical systems are represented by sets, properties by subsets, pure states by points. In particular, given two properties that are not mutually exclusive, one can always find states that possess both properties with certainty: The points that lie in the intersection of the corresponding sets. By contrast, the intersection of two non-orthogonal vector subspaces may be just one point: Thus, in quantum theory, given two properties, there may be no state that possesses both with certainty.

C.1.2 Tomography

The state ρ describes all the properties of the system. Everyone knows that the expectation value of any operator can be computed given the state. But the converse is also true: A sufficiently large set of expectation values must determine the state (otherwise, there would be information in ρ that no operator can capture). From the perspective of probability theory, tomography is the analog of reconstructing a probability distribution by sampling it—only here, one must reconstruct several distributions in order to reconstruct the state, because there exist incompatible measurements.

For instance, the state of *one qubit* can always be written

$$\rho = \frac{1}{2}(\mathbb{I} + \vec{m} \cdot \vec{\sigma}) \tag{C.1}$$

with $|\vec{m}| \leq 1$ and where $\vec{\sigma}$ are the Pauli matrices. So the state is fully known if the Bloch vector is. Now, the component of the Bloch vector along direction $\hat{n}$ is $\hat{n} \cdot \vec{m} = \mathrm{Tr}\left[\rho(\hat{n} \cdot \vec{\sigma})\right] \equiv \langle \hat{n} \cdot \vec{\sigma} \rangle$. Thus, in order to reconstruct ρ, it is enough to compute $\langle \sigma_{\hat{x}} \rangle$, $\langle \sigma_{\hat{y}} \rangle$ and $\langle \sigma_{\hat{z}} \rangle$.

Similarly, the state of *two qubits* can always be written as

$$\rho = \frac{1}{4}\left(\mathbb{I} + \vec{m}_A \cdot \vec{\sigma} \otimes \mathbb{I} + \mathbb{I} \otimes \vec{m}_B \cdot \vec{\sigma} + \sum_{ij=\hat{x},\hat{y},\hat{z}} T_{ij}\sigma_i \otimes \sigma_j\right) \tag{C.2}$$

and so it can be reconstructed by reconstructing the local Bloch vectors and the $\langle \sigma_i \otimes \sigma_j \rangle$.

C.1.3 Composite systems: Partial traces and purification

Given the state of a composite system, one can always infer the state of each of its subsystems. Let's consider bipartite systems for simplicity of notation: Given the state ρ_{AB}, we want to know which state ρ_A describes the observations made by Alice. In other words, we want

$$\mathrm{Tr}(\rho_A A) = \mathrm{Tr}[\rho_{AB}(A \otimes \mathbb{I}_B)] \tag{C.3}$$

for all operators A on Alice's system. Since $\mathbb{I}_B = \sum_{b=1}^{d_B} |\beta_b\rangle \langle \beta_b|$ where the $\{|\beta_b\rangle\}$ are an arbitrary basis of Bob's Hilbert space, one finds that the partial state is computed by a *partial trace*:

$$\rho_A = \mathrm{Tr}_B(\rho_{AB}). \tag{C.4}$$

States of composite systems are called *product states* if $\rho_{AB} = \rho_A \otimes \rho_B$. All the other states show *correlations*. Since the trace is basis-invariant, the kinematic correlations in quantum states are *no-signaling*: Whatever Bob does on his system, Alice will see the same ρ_A.

There is a complementary view to what we have just seen that is often useful. Consider a single system in a mixed state ρ: This state can be seen as the partial state of a composite system, and the composite system could be in a pure state. The construction of this *purification* is simple starting from the diagonal representation of ρ: If $\rho = \sum_i p_i |\psi_i\rangle \langle\psi_i|$, we have $\rho = \text{Tr}_B(|\Psi\rangle \langle\Psi|)$ with

$$|\Psi\rangle = \sum_i \sqrt{p_i} |\psi_i\rangle \otimes |\beta_i\rangle \tag{C.5}$$

where the $|\beta_i\rangle$ are mutually orthogonal. It can be proved that all purifications of a given state ρ are equivalent up to a local unitary on the purifying system (in other words, the only difference is which set of $|\beta_i\rangle$ is chosen).

C.2 Elementary Entanglement Theory

Entanglement theory was mostly developed in the late 1990s and the early decade of the twenty-first century. The Horodecki family made a commendable effort to compile the massive amount of knowledge generated during those days. Their review article (Horodecki *et al.*, 2009) is not what I would call pleasant reading, but is a unique resource for its comprehensiveness.

In this elementary survey, we give the definition of entanglement for an arbitrary number of subsystems, but focus on *bipartite* system for all the subsequent considerations.

C.2.1 Entangled states

A *pure state* of a composite system $\bigotimes_{n=1}^{N} \mathcal{H}_n$ is called entangled if it cannot be written as a product:

$$|\Psi\rangle \neq \bigotimes_{n=1}^{N} |\psi_n\rangle. \tag{C.6}$$

For *bipartite* pure states, there exist a canonical representation called *Schmidt decomposition*: There always exist two orthonormal bases $\{|\alpha_j\rangle | j = 1, \ldots, d_A\}$ and $\{|\beta_k\rangle | k = 1, \ldots, d_B\}$ such that

$$|\Psi\rangle = \sum_{i=1}^{\min(d_A, d_B)} \sqrt{p_i} |\alpha_i\rangle \otimes |\beta_i\rangle \tag{C.7}$$

with $p_i \geq 0$ for all i and $\sum_i p_i = 1$. For multipartite pure states, several generalizations have been studied, see e.g., (Acín *et al.*, 2001) for three qubits.

A *mixed state* is called entangled if there exist no convex decomposition over product states:

$$\rho \neq \sum_m p_m \left(\bigotimes_{n=1}^{N} \rho_{k,m} \right) \tag{C.8}$$

with $p_m \geq 0$ and $\sum_m p_m = 1$. A non-entangled state is called *separable*; often, a non-product separable state is called *classically-correlated state*.

While it is trivial to verify if a pure state is entangled or not, for a generic mixed state no algorithm is known. One can prove that a state is separable by exhibiting an explicit decomposition. As for proving that a state is entangled, we know only some sufficient criteria. The most used one is the *negativity of the partial transpose*. Let us illustrate it for two-qubit Werner states. The state reads (3.23)

$$\rho_W = \frac{1+W}{4}(|01\rangle\langle01| + |10\rangle\langle10|) + \frac{1-W}{4}(|00\rangle\langle00| + |11\rangle\langle11|)$$
$$- \frac{W}{2}(|01\rangle\langle10| + |10\rangle\langle01|).$$

If we transpose Bob's side of the matrix, we get

$$\rho_W^{T_B} = \frac{1+W}{4}(|01\rangle\langle01| + |10\rangle\langle10|) + \frac{1-W}{4}(|00\rangle\langle00| + |11\rangle\langle11|)$$
$$- \frac{W}{2}(|0\underline{0}\rangle\langle1\underline{1}| + |1\underline{1}\rangle\langle0\underline{0}|).$$

The smallest eigenvalue of $\rho_W^{T_B}$ is $\frac{1}{4}(1 - 3W)$, which is negative for $W > \frac{1}{3}$. Thus, every Werner state with $W > \frac{1}{3}$ is entangled.

For systems composed of two qubits, or of one qubit and one qutrit, the negativity of partial transpose is also necessary for entanglement. For any other composite systems, there exist mixed entangled states with positive partial transpose (PPT). These are usually called *bound entangled states* for a reason that will be mentioned in subsection C.2.3.

C.2.2   Witnessing entanglement

One may not have to reconstruct the full state before assessing whether entanglement is present. An *entanglement witness* is an operator W such that $\mathrm{Tr}(\rho W) \geq w_{\mathrm{sep}}$ for *all* separable states, while $\mathrm{Tr}(\rho W) < w_{\mathrm{sep}}$ for *some* entangled states. The analogy with Bell inequalities is evident: In fact, Bell operators are the only witnesses that detect entanglement in a device-independent way.

Here comes an example of an entanglement witness that is not a Bell operator. Consider two qubits. Given a product state $\frac{1}{4}(\mathbb{I} + \vec{m}_A \cdot \vec{\sigma}) \otimes (\mathbb{I} + \vec{m}_B \cdot \vec{\sigma})$, it is readily verified that $\langle\sigma_{\hat{x}} \otimes \sigma_{\hat{x}}\rangle + \langle\sigma_{\hat{y}} \otimes \sigma_{\hat{y}}\rangle + \langle\sigma_{\hat{z}} \otimes \sigma_{\hat{z}}\rangle = \vec{m}_A \cdot \vec{m}_B \leq 1$. Separable states are convex sums of product states: Therefore, for

$$W = \sigma_{\hat{x}} \otimes \sigma_{\hat{x}} + \sigma_{\hat{y}} \otimes \sigma_{\hat{y}} + \sigma_{\hat{z}} \otimes \sigma_{\hat{z}} \qquad\qquad (C.9)$$

it holds $-1 \leq \mathrm{Tr}(\rho W) \leq 1$ for all separable states. The singlet state has $\mathrm{Tr}(\rho W) = -3$: Thus, W detects its entanglement; and it is a simple exercise to change the signs in the expression of W to detect the the entanglement of other Bell states.

Now, how would W look like in a device-independent description? If the measurements are uncharacterized, all that we know is that there are three dichotomic measurements per player: So we'd have $W = A_0 B_0 + A_1 B_1 + A_2 B_2$, which can any value between -3 and $+3$ for LV. Thus, for W to be an entanglement witness, characterization is crucial: The system must be a qubit and the operators must be mutually unbiased Pauli matrices. For an example in which the operators can be uncharacterized, but one still needs the qubit assumption, see Appendix F.1.3.

C.2.3 Entanglement as a resource

The explosion in the study of entanglement theory was triggered by the realization that *entanglement is a resource* and the subsequent effort to quantify this resource. This quantification was phrased in terms of interconversion of resources. One first notices that entanglement cannot increase under local operations (LO), nor if classical communication (CC) among the players is allowed. Thus, one can ask: Having N copies of a given entangled state ρ, how many copies M of another entangled state $\rho\prime$ can I obtain by a processing that involves only LOCC?

In particular, taking the singlet state of two qubits as *unit of entanglement*, two basic tasks and the corresponding amount of entanglement can be defined:

- *Formation:* Having shared N_s two-qubit singlets, one wants to create M copies of a desired entangled state ρ. The entanglement of formation of ρ is then defined as $E_F(\rho) = \lim_{M\to\infty} \frac{N_s(M)}{M}$. Any entangled state has $E_F(\rho) > 0$.
- *Distillation:* Given N copies of an entangled state ρ, one wants to extract M_s two-qubit singlets. The entanglement of distillation (or distillable entanglement) of ρ is then defined as $E_D(\rho) = \lim_{N\to\infty} \frac{M_s(N)}{N}$. A state for which $E_D(\rho) > 0$ is called distillable.

For *pure* states, all quantifiers of bipartite entanglement coincide: $E = S(\rho_A) = S(\rho_B)$ where S is von Neumann entropy. In particular, the *maximally entangled states* of two qudits is one whose reduced states are $\rho_A = \rho_B = \frac{1}{d}\mathbb{I}_d$. By Schmidt decomposition, there exist a local basis in which these states read

$$|\Psi\rangle_{\mathrm{ME}} = \frac{1}{\sqrt{d}} \sum_{k=1}^{d} |k\rangle \otimes |k\rangle. \tag{C.10}$$

For *mixed* states, some cases aside, Murphy's law kicks in and everything that could go wrong does it. On the one hand, one does not have simple computable formulas because the limits cannot be resolved analytically. On the other hand, various quantifiers of entanglement do not coincide. The most striking example is the existence of states for which $E_F(\rho) > 0$ but $E_D(\rho) = 0$: Entanglement must be invested to create them, but no entanglement can be retrieved out of them. These non-distillable entangled states have been called *bound-entangled*. All PPT entangled states are bound-entangled;[3] whether some states with negative partial transpose (NPT) are also bound entangled remains open in spite of considerable effort.

C.3 Generalized Measurements or POVMs

C.3.1 Definition

The reader must be familiar with *projective measurements*, in which each output a is associated to a subspace of the Hilbert space of the system being measured, and therefore $P(a) = \mathrm{Tr}(\rho \Pi_a)$

---

[3] Indeed, it's not difficult to verify that LOCC cannot change PPT into NPT. But the singlet is obviously NPT: Therefore, it will be impossible to distill even one singlet from any number of copies of a PPT state under LOCC.

where Π_a is the projector on that subspace. There are two ways in which people have thought of generalizing this notion:

- At the algebraic level, one can think that a recipe like $P(a) = \mathrm{Tr}(\rho\Pi_a)$ produces valid probabilities for every state as long as $\Pi_a \geq 0$ for all a and $\sum_a \Pi_a = \mathbb{I}$. A set of orthogonal projectors satisfy these relations, but in addition they satisfy $\Pi_a^2 = \Pi_a$: This may not have to be enforced. This mathematical generalization explains why generalized measurements have been given the unfortunate, but by now standard, name of *positive-operator valued measures (POVMs)*.
- At the physical level: Upon receiving the system to be measured, one can append an auxiliary system[4] prepared in a given state, then perform a projective measurement on the joint system. Because the auxiliary system is in a given state known by the user, this procedure can be legitimately called a measurement on the system to be measured, since information gain will be only about that one.

Two definitions define the same class of measurements, an equivalence result known as *Naymark's theorem*: Any POVM can be seen as a projective measurement on an enlarged system; and given a projective measurement on an enlarged system, its effective description on the system to be measured is a POVM.

C.3.2   A semantic subtlety

In the text, we have often introduced operators that are linear combinations of POVM elements. For instance, in section 3.2 we have operators of the form $A = \Pi_{+1} - \Pi_{-1}$. These are convenient because $\mathrm{Tr}(\rho A)$ gives immediately the marginal bias $P(a = +1) - P(a = -1)$.

In a frequent abuse of language, one often takes these operators as the real thing and says "Alice measures A," certainly inspired by the very frequent quantum phrase "Alice measures $\sigma_{\hat{z}}$." As usual, imprecise language is fine as long as one knows its limits. Here I want to point out that the eigenvalues of A are $\{-1, +1\}$ only if the measurement is projective. Also, we need to recall that the labeling of the POVM outcomes is arbitrary: The same A could have been written $A = \Pi_0 - \Pi_1$, and then $\mathrm{Tr}(\rho A) = P(a = 0) - P(a = 1)$, while 0 and 1 have nothing to do with the eigenvalues of A even in the case of projective measurements.

C.3.3   POVMs and joint measurability

While examples of relevant POVMs are readily found in the literature, one issue that is usually tackled only by specialists is that of *joint measurability*. Since it was mentioned in the text, we have to introduce it. A widespread misunderstanding of quantum measurement theory alleges that one cannot measure position and momentum "at the same time." This is nonsense: One can obviously construct a box that outputs both a value for position and one for momentum: Only, the information that can be extracted will be less precise than what could be achieved by

---

[4] Auxiliary systems have been called *ancillae* for long time in the literature. The word means *servant maids* in Latin, and triggered some backlash recently.

measuring either position or momentum.[5] It is in this sense that position and momentum are *not jointly measurable*.

In general, projective measurements are jointly measurable if and only if they commute. But joint measurability is not so trivial for POVMs, to the point that a complete set of criteria for joint measurability is not known. For our goals, let us just look at the simplest example. Consider the two POVMS $\mathcal{M}_1 = \{\Pi_\pm^{(1)} = \frac{1}{2}(\mathbb{I} \pm \eta\sigma_z)\}$, $\mathcal{M}_2 = \{\Pi_\pm^{(2)} = \frac{1}{2}(\mathbb{I} \pm \eta\sigma_x)\}$. Obviously, the elements of $\mathcal{M}_1$ do not commute with those of $\mathcal{M}_2$ as soon as $\eta > 0$. Nonetheless, for $\eta \leq \frac{1}{\sqrt{2}}$ they are jointly measurable. Indeed, consider the four-output POVM defined by

$$\Pi_{+,+} = \frac{1}{4}\left(\mathbb{I} + \frac{\sigma_z + \sigma_x}{\sqrt{2}}\right), \quad \Pi_{-,-} = \frac{1}{4}\left(\mathbb{I} - \frac{\sigma_z + \sigma_x}{\sqrt{2}}\right)$$

$$\Pi_{+,-} = \frac{1}{4}\left(\mathbb{I} + \frac{\sigma_z - \sigma_x}{\sqrt{2}}\right), \quad \Pi_{-,+} = \frac{1}{4}\left(\mathbb{I} - \frac{\sigma_z - \sigma_x}{\sqrt{2}}\right).$$

It holds $\{\Pi_a^{(1)} = \Pi_{a,+} + \Pi_{a,-}$ and $\{\Pi_a^{(2)} = \Pi_{+,a} + \Pi_{-,a}$ for $\eta = \frac{1}{\sqrt{2}}$. For smaller values of η, one can just add noise to the four-output apparatus. In other words, one can have a single measurement device with four outputs, such that a clever grouping of the outputs reproduces $\mathcal{M}_1$ and $\mathcal{M}_2$ exactly.

---

[5] Here is a very simple construction: Hide in the measurement device a box that measures position and one that measures momentum; in each round, the device tosses an unbiased coin and routes the physical system to one of the boxes, while instructing the other box to output something. It does not look optimal and indeed it is not; but it does extract some information about position and momentum, and is thus a legitimate measurement device for both. There exist bounds that quantify the loss of information independently of the constructions, usually going under the name of *Arthurs-Kelly uncertainty relations*.

Appendix D
LV Models for Single Systems

Knowing that LV models fail to describe observed phenomena, it may look pointless to scrutinize them in further detail. Still, on the side of foundations of physics, a sizeable amount of work has been devoted precisely to that. The reason is that, from the perspective of interpreting quantum theory, Bell nonlocality is not fully satisfactory, insofar as it treats single systems in a trivial way. It is more than fair to argue that non-classicality should be captured by a notion that applies to single systems too.

In this appendix, I mainly review notions that lead to the Kochen-Specker version of contextuality and its recent improvements, because these are the notions that are most closely related to nonlocality. A text more foundational than mine should devote a sizeable section to the discussion of so-called ψ-epistemic models and their restrictions, triggered by the result of Pusey, Barrett, and Rudolph (2012). For these developments, I direct the reader to the masterful review by Matthew Leifer (2014).

D.1   Overview: Looking into Measurements

Let us start from Fine's theorem (subsection 2.3.3): A behavior is local if and only if there exist a joint probability distribution for the outputs. For a single system, this is clearly the case: For every input, one can specify the outputs, so one could just take $P(a_1, a_2, \ldots, a_m) = \prod_{x=1}^{m} P(a|x)$. But in physics, to the input x one associates a measurement $\mathcal{M}^x = \{M_a^x | a \in \mathcal{A}\}$. The paths to discovering quantum effects even in single system pass through *looking more closely into those measurements* (at the price of losing the device-independence that is the power of Bell nonlocality).

The obvious path is the one through the full characterization of the degrees of freedom that are being measured. If I know that what is being measured is a magnetic moment, and that the apparatus is a gradient of magnetic field, then no classical magnetic moment will behave as observed in the Stern-Gerlach experiment. If I know that homodyne measurements are being performed on optical fields, then no classical field can yield a negative Wigner function. The list can obviously continue to include all "quantum effects" reported in the last century, and those which mostly concern single systems without any hint of Bell nonlocality. We refer to your favorite list of such phenomena for details.

But one can see issues with LV descriptions with a much weaker level of characterization. Each of the M_a^x represents a physical property. Suppose now that two *different* measurements $\mathcal{M}^{\bar{x}} \neq \mathcal{M}^{\bar{x}'}$ are guaranteed to share one physical property, say $M_1^{\bar{x}} = M_2^{\bar{x}'}$. A classical theory would then request that $a_{\bar{x}} = 1$ if and only if $a_{\bar{x}'} = 2$: Either the system possesses that property, or it doesn't, prior to the measurement and irrespective of which other properties are tested alongside with it (the "context"). Kochen and Specker (1967) proved that, if one takes enough many measurements that share some common properties, it is impossible to assign outputs according to the classical recipe. In other words, a LV model can assign a determined output to every measurement taken as a whole, but can't assign pre-existing physical properties. This is the starting point of *contextuality*. Before delving more into it (section D.3), we review the only physical system for which the KS theorem does *not* apply.

D.2 Bell's Local Hidden Variable Model for One Qubit

A LV model that reproduces exactly the quantum statistics of *projective measurements on a single qubit* was provided by Bell (1966). If the qubit is in the state $\frac{1}{2}(1 + \vec{m} \cdot \vec{\sigma})$, the quantum prediction for measurement along direction $\hat{a}$ is

$$P(a|\hat{a}) = \frac{1}{2}(1 + a\vec{m} \cdot \hat{a}), \text{ that is } \langle a \rangle_{\hat{a}} = \vec{m} \cdot \hat{a} \tag{D.1}$$

for $a \in \{-1, +1\}$. In the LV model, the state of the system is represented by $(\vec{m}, \vec{\lambda})$ where $\vec{\lambda}$ is a unit vector. The output is deterministically computed to be

$$a(\vec{\lambda}) = \text{sign}\left[(\vec{m} - \vec{\lambda}) \cdot \hat{a}\right] \tag{D.2}$$

which is either $+1$ or -1 as it should.

Assuming that $\vec{\lambda}$ is drawn from the uniform distribution on the sphere, i.e., $\rho(\vec{\lambda})d\vec{\lambda} = \frac{1}{4\pi}\sin\theta d\theta d\varphi$, it is easy to prove[1] that

$$\langle a \rangle_{\hat{a}} = \int d\vec{\lambda}\rho(\vec{\lambda})a(\vec{\lambda}) = \vec{m} \cdot \hat{a} \tag{D.3}$$

which matches the quantum prediction. Notice that nothing in this model requires $\vec{\lambda}$ to be drawn "at random" in each round: The sequence of $\vec{\lambda}$ may be pre-registered.

Bell's model has all the desired features of a LV model. It is not "contextual" for a simple reason: The property $\vec{m}$ can only be measured together with $-\vec{m}$, and thus there is only one "context." Even so, one can find it unpleasant for its "ontological excess baggage" (Hardy, 2004): While each measurement can extract at most one bit, the model describes the system by a vector i.e., with an infinite amount of information–and it was proved that one cannot get away with less (Montina, 2006; Dakić *et al.*, 2008).

D.3 Contextuality

D.3.1 History at a glance

The mathematical definition of Bell nonlocality is peacefully accepted by everyone. By contrast, one finds various formal definitions of contextuality. At the moment of writing, two main threads can be identified.

---

[1] Without loss of generality, one can choose $\hat{a} = \hat{z} \equiv (\theta = 0, \varphi)$, since nothing else in the problem specifies the choice of spherical coordinates.

The first thread stems from the pioneering work of Kochen and Specker (KS) (1967). As sketched previously, it is an attempt of attributing truth values to physical properties, reaching the conclusion that this is not possible. Over the years, the argument underwent several simplifications, but retained assumptions that are considered problematic: It works only for *deterministic* hidden variables, it requires importing structures typical of quantum theory, and it is based on the possibility of performing measurements that are "ideal" under several respects. In 2008 Klyachko, Can, Binicioglu, and Shumovsky (KCBS) managed to remove the assumptions of determinism and the reference to quantum structures in KS-type contextuality (Klyachko *et al.*, 2008), leading to a revival of results.

The second thread started a few years before the KCBS breakthrough, when Spekkens (2005) proposed a radically new approach to address non-classicality of single systems, which he also called contextuality. His approach relies on the formalism of ontological models, that have been the object of several works in foundations (Leifer, 2014). Instead of trying and attributing truth values to properties for any single measurement, Spekkens reasons on the structure of theory: He asks that two preparations that give the same statistics for all measurements should be identified ("preparation non-contextuality"), and that two measurements that give the same statistics for all preparations should be identified ("measurement non-contextuality"). This approach gets away from the need of defining ideal measurements, although the experimental difficulty is only displaced: One must now be convinced that two preparations (or two measurements) are identical. Intriguingly, measurement contextuality can be demonstrated also for one qubit. This is welcome, insofar as we all believe that a qubit is a quantum system too; but it shows how deep into theoretical structures this definition must delve, in order to find flaws even in the LV model presented in section D.2.

The Spekkens' approach definitely belongs to the important advances in quantum foundations, but it is too far from Bell nonlocality to be discussed in this appendix. In the remainder of this section, we shall only introduce the KS-type approaches, starting from KCBS.

D.3.2   The KCBS inequality

We present the KCBS construction. In a device-independent perspective, it consists of a single device with five inputs and four outputs. In order to uncover non-classicality, we need to add some structure. We assume that each input calls two processes, each with binary output. Specifically, we consider five processes A_k, $k \in \{1, 2, 3, 4, 5\}$, and the inputs call the pairs $\{(1, 2), (2, 3), (3, 4), (4, 5), (5, 1)\}$. The output of the process A_j is denoted $a_j \in \{-1, +1\}$; the output of the device to the input (j, k) is therefore the pair (a_j, a_k). The need for "ideal" measurements manifests itself as follows: The process A_j must be the same, whether it is called together with A_{j-1} or with A_{j+1}.

At this point, we proceed as we did when we introduced CHSH (section 1.3): We assume that all $(a_1, a_2, a_3, a_4, a_5)$ have a well-defined value. Then it is easy to prove that

$$a_1 a_2 + a_2 a_3 + a_3 a_4 + a_4 a_5 + a_5 a_1 \geq -3. \tag{D.4}$$

This leads to the corresponding inequality for the observed average values. KCBS proved that there exist a set of five binary observables on a qutrit that commute pairwise, and such that the inequality is violated for some quantum states.

D.3.3   Before and after KCBS

Having understood KCBS, we can appreciate the difference with the original KS proof, as well as some later improvements.

The original KS proof relied on the fact that projective measurements on qutrits are defined by three orthogonal projectors. In each such triple, a non-contextual LV model would assign one 1 (for the property that the system possesses) and two 0's. The KS *tour de force* consisted in identifying 132 such triples, involving 117 projectors, such that no such assignment can be made: There will always be a triple that has either three 0's or more than one 1. The geometry imposed by the structure of quantum theory plays a crucial role, and the argument fails if one does not request the LV model to be deterministic. Both limitations are absent in the KCBS argument: Only the notion of joint measurability is retained, and an analog of Fine's theorem can be proved to connect deterministic and non-deterministic LVs.

One very striking feature of the original KS proof is that it is *state-independent*: The contradiction is due to the quantum structure of the measurements, irrespective of any description of the state. This is very different from what happens in Bell nonlocality: There, LV models can always be constructed, it's only some observed statistics (some states) that are incompatible with them. The KCBS inequality is state-dependent, but it took only a few months for Cabello to find a similar inequality that is state-independent (Cabello, 2008). A few years later, Yu and Oh (2012) found a state-independent criterion that requires only 13 projectors; and this was proved to be the smallest possible number (Cabello *et al.*, 2016).

D.3.4   Contextuality and nonlocality: A comparison

We conclude this appendix with a comparison between contextuality and nonlocality. It is often said that contextuality is "more general": My experience in science is that any statement containing this g-word needs qualification.

Here are the similarities:

- In terms of mathematical description, one can indeed say that nonlocality is a special case of contextuality. Both rely on versions of Fine's theorem and the corresponding possibility of enclosing the classical options in a polytope. Within quantum theory, a scenario is defined by listing which operators commute (see the definition Q' of the quantum set in subsection 6.2.2). One of the most explicit effort to show the common structure of nonlocality and contextuality is the work of Cabello, Severini, and Winter (2010, 2014).

- In terms of physical principles, as we have mentioned in section 10.4, the physical principle that was called Local Orthogonality for nonlocality has its exact counterpart for contextuality, where it was called Consistent Exclusivity (and where it seems to be actually more powerful in singling out quantum behaviors).

However, there are also important differences.

- Within quantum theory, contextuality can be made state-independent, while nonlocality can't. Also, the commutation relations for nonlocality are imposed by the locality condition (and in this sense can be called "device-independent"). By contrast, in a generic scenario of contextuality, and notably in single-particle scenarios, the commutation relations must be justified by some characterization of the devices.

- At a more operational level, it is interesting to compare the *resources needed for the classical simulation* of such non-classical behaviors. The classical simulation of Bell nonlocality requires *communication*, as abundantly discussed in this book. For some time, it was believed that the simulation of contextuality would be trivial, but recently it was noticed that the simulation of some contextuality scenarios require *memory* (Tavakoli and Cabello, 2018). Now, one can try a unifying narrative by noticing that communication is "transmission of information across space," memory is "transmission of information across time." But our current understanding of physics breaks this symmetry. Because communication faster than light is impossible, one can exclude communication by arranging space-like separated events. By contrast, memory is only limited by storage capacity (the amount of information that can be stored and the lifetime of the storage). While surely finite, these capacities are very large and constantly being improved: Only very complex contextuality criteria may possibly challenge the current state-of-the-art. Notice that one could base nonlocality on similar considerations: One could believe that communication may happen at any speed, then base the evidence of nonlocality on Bell tests different than those we described, whose simulation involves very high communication complexity. While these definitely exist on paper (subsection 11.2.2), none is practically feasible.

Let me finish with a more subjective argument for the difference between communication and memory. Devices do not communicate spontaneously, they must have been engineered on purpose; by contrast, a device may retain information about its past even without being purposefully designed for it. Similarly, we could easily believe that a physical system may retain some memory of its past states; it's far less easy to believe that two devices communicate if we don't see (with our eyes, or with our knowledge of physics) a communication channel.

Appendix E
Basic Notions of Convex Optimization

This appendix begins with a very concise presentation of the main notions of convex optimization. This has been distilled from (Boyd and Vandenberghe, 2004), keeping the same notations so that the readers can easily continue their study with that book. Two examples introduced in the main text are then worked out explicitly.

E.1 Generalities

E.1.1 Convex programs and their Lagrange dual

A *convex program* is an optimization that can be cast in the following form:

$$
\begin{aligned}
p^* = \min f_0(x) \\
\text{subject to} \quad f_i(x) \leq 0, i = 1, \dots, m \\
h_i(x) = 0, i = 1, \dots, p
\end{aligned}
\tag{E.1}
$$

where $x \in \mathbb{R}^n$ are the variables, the objective function f_0 and the functions f_i are *convex* functions with value in $\mathbb{R}$, and the h_i are *affine* functions with value in $\mathbb{R}$. A point x in the intersection $\mathcal{D}$ of the domains of these functions is called *feasible* if it satisfies the constraints. If no point satisfies the constraints, the program is infeasible and by convention one sets $p^* = +\infty$.

Being an optimization under constraints, it is natural to approach it with the techniques of Lagrange multipliers. The Lagrangian is defined as the function $\mathcal{D} \times \mathbb{R}_+^m \times \mathbb{R}^p \longrightarrow \mathbb{R}$

$$
L(x, \lambda, v) = f_0(x) + \sum_{i=1}^{m} \lambda_i f_i(x) + \sum_{i=1}^{p} v_i h_i(x).
\tag{E.2}
$$

From this, one defines the Lagrange dual function

$$
g(\lambda, v) = \inf_{x \in \mathcal{D}} L(x, \lambda, v).
\tag{E.3}
$$

Because of the convexity assumptions in the primal and the positivity of λ, $g(\lambda, v)$ is concave and satisfies $g(\lambda, v) \leq p^*$. The *Lagrange dual program of the primal (E.1)* is

$$
d^* = \sup_{\lambda \in \mathbb{R}_+^m, v \in \mathbb{R}^p} g(\lambda, v).
\tag{E.4}
$$

It is also a convex problem, since it can be rewritten as the minimization of the convex function $-g$. Notice that this problem is formally unconstrained; however, natural constraints may appear, as we shall see in the next example of linear programs.

Clearly, $d^* \leq p^*$. When equality holds, one says that the convex program exhibits *strong duality*; when it does not hold, the quantity $p^* - d^*$ is called *duality gap*.

As mentioned in the main text, the possibility of writing a dual program allows to compute both an upper and a lower bound for the quantity of interest (typically the solution of the primal). Let us make this statement more precise. If both the primal and the dual are feasible, for any feasible x_f of the primal and any feasible $\xi_f \equiv (\lambda, \nu)_f$ of the dual it holds

$$g(\xi_f) \leq d^* \leq p^* \leq f_0(x_f). \tag{E.5}$$

If in addition strong duality holds, the algorithm usually converges to these bounds being the same within numerical precision.

E.1.2   Special case: Linear programs

A convex program is a *linear program* if all the functions f_0 and f_i are affine. The objective function can then be taken as linear, since a constant can always be added after the optimization. The very frequent case $f_i(x) = -x_i$ defines the *standard form* of a linear program

$$\begin{aligned} p^* = &\min_x c^T x \\ \text{subject to} \quad &x \geq 0 \\ &Ax = b \end{aligned} \tag{E.6}$$

A typical example is the optimization over a stochastic vector, $x \geq 0$ imposing the non-negativity of probabilities. The membership problem for the local polytope (2.22) can be cast in this form, see subsection E.2.1.

Let us find the Lagrange dual of (E.6). The Lagrangian is

$$L(x, \lambda, \nu) = c^T x - \lambda^T x + \nu^T (Ax - b) = -\nu^T b + (c - \lambda + A^T \nu)^T x.$$

The Lagrange dual function is the infimum over x, which is $-\infty$ unless $c - \lambda + A^T \nu = 0$. We can treat this condition as a constraint for the dual, since it is a necessary condition to obtain a non-trivial bound; and we can rewrite it as $A^T \nu + c \geq 0$ since $\lambda \geq 0$. In summary, the Lagrange dual program of (E.6) is

$$\begin{aligned} d^* = &\max_\nu -b^T \nu \\ \text{subject to} \quad &A^T \nu + c \geq 0. \end{aligned} \tag{E.7}$$

E.1.3   Special case: Semidefinite programs (SDPs)

A *semidefinite program* differs from a linear program in that the affine *inequality* constraints read $\sum_i x_i F_i \leq 0$ where $F_i \in \mathbb{R}^{k \times k}$ are symmetric matrices of some size (the case of linear program is recovered for $k = 1$). So the primal is given by

$$\begin{aligned} p^* = &\min_x c^T x \\ \text{subject to} \quad &\sum_i x_i F_i \leq 0. \\ &Ax = b \end{aligned} \tag{E.8}$$

The Lagrangian is similar as before, but now the Lagrange multiplier of the semi-definite constraint is a matrix $Z \geq 0$:

$$L(x, \lambda, \nu) = c^T x + \mathrm{Tr}(Z \sum_i x_i F_i) + \nu^T (Ax - b)$$

$$= -\nu^T b + \sum_i (c_i + \mathrm{Tr}(ZF_i) + (A^T \nu)_i) x_i.$$

The same reasoning as previously demonstrated implies that the Lagrange dual is

$$
\begin{aligned}
d^* &= \max_{\nu, Z} -b^T \nu \\
\text{subject to} \quad &c_i + \mathrm{Tr}(ZF_i) + \sum_j A_{ji} \nu_j = 0 \text{ for all } i \\
&Z \geq 0
\end{aligned}
\tag{E.9}
$$

E.2   Two Explicit Examples

E.2.1   The membership problem for the local polytope

As an example of linear program, we look at the membership problem for the local polytope (2.22). The variable is $x = \underline{q} \in \mathbb{R}^{\sharp_{LD}}$ the vector of weights of the extremal points of the local polytope. In the formulation of that problem in the main text, there is no objective function to minimize. One could be tempted to minimize $|\mathcal{P} - \sum_{i=1}^{\sharp_{LD}} q_i \mathcal{P}_{LD,i}|$, which is not linear but there exist tricks to write a linear program, see Eqs (1.6) and (1.7) of (Boyd and Vandenberghe, 2004). But the standard solution is simpler: One takes all the conditions (2.22) as constraints and sets $f_0(x) = 0$. The outcome will then be $p^* = 0$ if the program is feasible, $p^* = +\infty$ if it is not—and this is exactly what we want to know: Whether the constraints can be satisfied or not.

Because the only inequality constraint is the non-negativity of probabilities $x \geq 0$, and all the equality constraints are affine, this problem can be cast as a linear program in the standard form (E.6). Clearly, $c = 0 \in \mathbb{R}^{\sharp_{LD}}$. To identify A and b, recall that a behavior can be represented by a vector in $\mathbb{R}^{D_{NS}}$ (for instance, the entries of the Collins-Gisin representation). Let us denote by $v_{\mathcal{P}}$ such a vector, using for simplicity v_i for the vector representing the local deterministic behavior $\mathcal{P}_{LD,i}$. Then $A \in \mathbb{R}^{(D_{NS}+1) \times \sharp_{LD}}$ and $b \in \mathbb{R}^{D_{NS}+1}$ are given by

$$
A = \begin{pmatrix} v_1 & v_2 & \cdots & v_{\sharp_{LD}} \\ 1 & 1 & \cdots & 1 \end{pmatrix}, \quad b = \begin{pmatrix} v_{\mathcal{P}} \\ 1 \end{pmatrix},
\tag{E.10}
$$

the last entries describing the constraint $\sum_i q_i = 1$.

Let us now turn to the description (E.7) of the dual. Denoting by $v\prime$ the first D_{NS} components of v, and by v_0 the last component, it's easy to verify that the dual reads

$$
\begin{aligned}
d^* &= \max -(v_{\mathcal{P}}^T v' + v_0) \\
\text{subject to} \quad &v_k^T v' + v_0 \geq 0 \forall k = 1, \ldots, \sharp_{LD}.
\end{aligned}
\tag{E.11}
$$

If $\mathcal{P}$ is local, $v_{\mathcal{P}}$ is a linear combination of the v_k; then the constraints force $v_{\mathcal{P}}^T v' + v_0 \geq 0$, whence $d^* = 0$. Thus, if the primal is feasible, the dual is also feasible and we have strong duality. Conversely, if $\mathcal{P}$ is nonlocal, one can find v such that $v_{\mathcal{P}}^T v' + v_0 < 0$. This is nothing else than the definition of a Bell inequality violated by $\mathcal{P}$. Once such a v is found, any scaling αv is also a solution; therefore $d^* = +\infty$. Thus, when the primal is infeasible, the dual is unbounded. In order for the algorithm to output a Bell inequality with finite coefficients, one can add the additional constraint $v_{\mathcal{P}}^T v' + v_0 \geq -1$ to the dual, as suggested in Eq. (19) of (Brunner *et al.*, 2014).

E.2.2   An example of SDP

As an example of SDP, we look at a relaxation of the quantum upper bound for the guessing probability (8.8) that reads

$$\bar{P} = \max_{\{\mathcal{P}_{(a',b')}\}} \sum_{(a',b')} P_{(a',b')}(a',b'|0,0)$$
$$\text{subject to} \quad \sum_{(a',b')} \mathcal{P}_{(a',b')} = \mathcal{P} \qquad . \tag{E.12}$$
$$\mathcal{P}_{(a',b')} \in \mathcal{Q}_{\mathcal{F}} \text{ for all } a' \in \mathcal{A}, b' \in \mathcal{B}$$

The last condition is captured by the positivity of hermitian moment matrices $\Gamma_{(a,b)}$. Now we are going to cast this problem in the standard form, following (Bancal *et al.*, 2014). To avoid confusion, the inputs of the Bell test will be denoted by X, Y, leaving x, y for the semi-definite variables.

Let's set $\Gamma_{(a',b')} = \sum_i x_{(a',b')}^i F_i$ where $\{F_i\}$ is a fixed basis of hermitian matrices. Let's also define K_{abXY} as the constant matrix that picks up the element $(a,b|X,Y)$ of a behavior in the moment matrix, that is $P_{(a',b')}(a,b|X,Y) = \text{Tr}(K_{abXY}\Gamma_{(a',b')}) = \sum_i c_{abXY}^i x_{(a',b')}^i$ with $c_{abXY}^i = \text{Tr}(K_{abXY}F_i)$. Now the primal reads

$$\bar{P} = \max_x \sum_{(a',b')} \sum_i c_{a'b'00}^i x_{(a',b')}^i$$
$$\text{subject to} \quad \sum_{(a',b')} \sum_i c_{abXY}^i x_{(a',b')}^i = P(a,b|X,Y) \text{ for all } a,b,x,y. \tag{E.13}$$
$$\sum_i x_{(a',b')}^i F_i \geq 0 \text{ for all } a',b'$$

This is of the form (E.8) with the opposite sign for x, the matrix $A_{ji} = c_{abXY}^i b_j = -P(a,b|X,Y)$ with $j \equiv (abXY)$. Thus we know that the dual will read (using $y = -v$)

$$\bar{d} = \min_{y,Z} \sum_{abXY} P(a,b|X,Y)y_{abXY}$$
$$\text{subject to} \quad c_{a'b'00}^i + \text{Tr}(F_i Z_{(a',b')}) - \sum_{abXY} c_{abXY}^i y_{abXY} = 0 \text{ for all } a',b',i. \tag{E.14}$$
$$Z_{(a',b')} \geq 0 \text{ for all } a',b'.$$

If strong duality holds, i.e., $\bar{d} = \bar{P}$, the solution y^* of the dual defines a hyperplane in the space of behaviors, such that $\sum_{abXY} P(a,b|X,Y)y_{abXY}^* = \bar{P}$. In other words, the result of the optimization can be read by evaluating the suitable Bell inequality on the observed behavior (though this suitable inequality usually can't be guessed without solving the problem anyway).

Appendix F
Device-Independent Certification: History and Review

F.1 Bell Nonlocality and Quantum Information Science

F.1.1 Before 2005: Nonlocality as an inspiration from the past

Quantum information science burst on the scene of physics in the last decade of the twentieth century. Quantum cryptography (more precisely, *quantum key distribution*, QKD) was invented by Bennett and Brassard (1984) and by Ekert (1991). It promised the possibility of distributing a secret between distant players communicating over a channel that can be tapped into, something impossible to achieve without the exchange of quantum systems. Shortly after that, Shor (1994) proved that a *quantum computer* would reduce the time needed to solve some specific problems, notably the factoring of large integers into prime factors.

The work of Bell has constantly been cited as an *inspiration* for quantum information science, insofar as it had replaced philosophical debates with a testable argument. The quantum information attitude was similar. Take the example of incompatible measurements: Generations of physicists have been puzzled by this fact, and many had even hoped to find a way out of that. Bennett and Brassard, on the contrary, accept the fact and use it to fool an eavesdropper: If the infamous Eve taps on the channel to learn the secret information carried by the quantum system, she will unavoidably modify the state of the system and her presence will be noticed. Ekert used Bell nonlocality itself to reach the same conclusion: If the outputs were not determined prior to the measurements, as the violation of a Bell inequality certifies, then those correlated numbers are secret for anyone that has not seen them. Indeed, nobody could have a copy of numbers that did not exist.

However, for several years Bell nonlocality was *not* a major topic in quantum information science. Sure enough, riding on the success and popularity of entanglement theory, the theory of nonlocality also thrived more than ever before: Many of the works cited in Part I were produced during those years, and nonlocality was brought into the community of theoretical computer science through the formalization of "non-local games" (Cleve *et al.*, 2004). Also, some experimental groups took advantage of the favorable climate to realize improved Bell tests, whose intriguing character was duly noted (Weihs *et al.*, 1998; Tittel *et al.*, 1998; Stefanov *et al.*, 2002). But from a quantum information perspective, Bell inequalities were seen as *sub-optimal entanglement witnesses* that trade power in detection of entanglement (they will never detect the entanglement of Werner states whose state behavior is local) with the capacity to rule out LV models (already very convincingly falsified by several experiments). By and large, the future seemed to belong to approaches better tailored to the advances of entanglement theory.

The change of perspective took place around the year 2005 and was completed by 2007, opening the field of device-independent (DI) certification. At the moment of writing DI certification has grown into a full sub-field of quantum information (section F.2). This development has not come without confusions, notably about the actual extent of the independence from the devices. These will be clarified in section F.3. Before going to that, we revisit the history of device-independence.

F.1.2   The tortuous path to device-independence

Pre-history first—or lack thereof: The notion of device-independence remains hidden in all the classic literature on nonlocality. This is not to say that derivations of Bell inequalities were wrong: They are certainly correct in all the works cited in Part I and many others; a close scrutiny may even find traces of the notion, which is natural after all. But certainly no prominence was given to it, rather the opposite: Emphasis was often put on the physical degree of freedom for which some inequalities were thought to be tailored.[1]

The idea that Bell inequalities can be used for *certification* appears first and very explicitly in Artur Ekert's re-discovery of QKD (Ekert, 1991). In one of those strange turns of fate, the intuition was immediately *killed*: Bennett, Brassard, and Mermin (1992) set out to argue that the violation of Bell's inequality is not the right way to look at QKD, one has to look for entanglement.[2] They went on to prove that, for qubits, Ekert's protocol is equivalent to an entanglement-based version of the original BB84 protocol. The *caveat* "for qubits" was not noticed, by them or anyone else: The QKD literature uniformly repeated the claim that Ekert's protocol and BBM92 are equivalent till 2006.

Device-independence in QKD resurfaced in the pioneering work of Mayers and Yao (1998; 2004) that we have presented in detail in section 7.2. Mayers and Yao only proved self-testing of the ideal correlations, which doesn't count as a security proof—to their credit, unconditional security proofs had not been developed yet: Mayers himself was finalizing the first one. To add to the complication, Mayers' papers were very hard to read because of their mathematical sophistication. In this situation, the Mayers-Yao result was taken note of but not followed up: The field of QKD rather grew using more understandable approaches to security proofs.

The work that proved decisive to trigger the awareness of device-independence was to come again from the side of key distribution, but with a very foundational twist. Barrett, Hardy, and Kent (2005*a*) realized that one could prove security using only the no-signaling assumption based on nonlocality. They invented a suitable protocol and provided a proof of principle, in which one secure bit can be extracted from an infinitely long raw key. In the last paragraph, this paper comments on Ekert's protocol—but, quite astonishingly, upholds the received knowledge that nonlocality does not really play a role there.

---

[1]   As already mentioned, Bell inequalities with d outputs were routinely referred to as being "for qudits" even in the title of papers (Collins *et al.*, 2002*a*; Kaszlikowski *et al.*, 2002; Żukowski and Brukner, 2002). For another example among many, Belinsky and Klyshko (1993*a*; 1993*b*) present their derivation of the MABK family as novel because they had optical interference experiments in mind, while previous works described it with spins. As for textbooks of quantum mechanics or even quantum information that mention Bell nonlocality, the presentation always considers a concrete implementation of Bell tests, either with spins of electrons (Bohm's thought experiment) or with polarization of photons (Aspect's real experiment). This element of concreteness is certainly welcome for beginners, but the prose all too often mixes the structure of the Bell test with technical considerations about the devices.

[2]   In the previous subsection I mentioned the marginalization of nonlocality by the emergent entanglement theory: It was definitely based on personal recollections, but this paper is an objective witness of that atmosphere.

Nevertheless, the idea was launched. In a paper that elaborates on security under the sole no-signaling assumption, Acín, Gisin, and Masanes (2006) finally stressed that the equivalence between BB84 and Ekert holds only for qubits.[3] This crucial insight is presented in detail in the next subsection.

F.1.3 The last step

The behavior that describes the correlations expected in an ideal, error-free run of the Bennett-Brassard 1984 (BB84) protocol for QKD is

$$P(a, b|x, y) = \frac{1}{4}(1 + ab\delta_{x,y}) \tag{F.1}$$

with $x, y \in \{0, 1\}$ and $a, b \in \{-1, +1\}$. In other words: When $x = y$ we have only $(a, b) = (+1, +1)$, or $(-1, -1)$ with equal probability; when $x \neq y$, a, and b are uncorrelated.

Let's first assume Alice's and Bob's systems are qubits. A two-output POVM on a qubit $\{E_+, E_-\}$ can be parametrized as $E_c = \frac{1}{2}(\gamma_c + c\eta\hat{c} \cdot \vec{\sigma})$ with $\gamma_c \geq 0$, $\gamma_+ + \gamma_- = 2$, $0 \leq \eta \leq \min(\gamma_+, \gamma_-)$, and $|\hat{c}| = 1$. Using such expressions for Alice's and Bob's measurements and the generic form of a two-qubit state (3.12), one obtains for a generic behavior

$$P(a, b|x, y) = \frac{1}{4}\left[\gamma_a^x \gamma_b^y + a\eta_x \gamma_b^y \text{Tr}(\rho\hat{a}_x \cdot \vec{\sigma}) + b\eta_y \gamma_a^x \text{Tr}(\rho\hat{b}_y \cdot \vec{\sigma}) + ab\eta_x \eta_y T_{xy}\right]$$

with $T_{xy} = \text{Tr}(\rho\hat{a}_x \cdot \vec{\sigma} \otimes \hat{b}_y \cdot \vec{\sigma})$. To recover (F.1) it is necessary to set $\gamma_a^x = \gamma_b^y = 1$ for all (a, b, x, y); also, since $|T_{xy}| \leq 1$, we need $\eta_x = \eta_y = 1$ for all (x, y): That is, the measurements are projective. Furthermore, the state and measurement directions must satisfy

$$\text{Tr}(\rho\hat{a}_x \cdot \vec{\sigma}) = \text{Tr}(\rho\hat{b}_y \cdot \vec{\sigma}) = 0, \tag{F.2}$$

$$\text{Tr}(\rho\hat{a}_x \cdot \vec{\sigma} \otimes \hat{b}_y \cdot \vec{\sigma}) = \delta_{x,y}. \tag{F.3}$$

Without loss of generality, we can set $\hat{a}_0 = \hat{b}_0 = \hat{z}$, $\hat{a}_1 = \cos\alpha\hat{x} + \sin\alpha\hat{z}$ and $\hat{b}_1 = \cos\beta\hat{x} + \sin\beta\hat{z}$. The conditions (F.3) read $T_{\hat{z}\hat{z}} = 1$, $T_{\hat{x}\hat{z}} = -\tan\alpha$, $T_{\hat{z}\hat{x}} = -\tan\beta$ and $T_{\hat{x}\hat{x}} = \frac{1+\sin\alpha\sin\beta}{\cos\alpha\cos\beta}$. However, $T_{\hat{z}\hat{z}} = 1$ forces $T_{\hat{x}\hat{z}} = T_{\hat{z}\hat{x}} = 0$. Indeed, consider $T_\theta \equiv \text{Tr}[\rho\sigma_{\hat{z}} \otimes (\cos\theta\sigma_{\hat{x}} + \sin\theta\sigma_{\hat{z}})]$. On the one hand, $|T_\theta| \leq 1$ must hold because it's the expectation value of an operator whose eigenvalues are $+1$, 0, and -1. One the other hand, $T_\theta = \cos\theta T_{\hat{z}\hat{x}} + \sin\theta T_{\hat{z}\hat{z}}$, whose achievable maximum over θ is $\sqrt{T_{\hat{z}\hat{x}}^2 + T_{\hat{z}\hat{z}}^2}$. So, if $T_{\hat{z}\hat{z}} = 1$, $T_{\hat{z}\hat{x}}$ must be equal to zero. The proof is identical for $T_{\hat{x}\hat{z}}$, changing the roles of Alice and Bob. Thus we are forced to set $\sin\alpha = \sin\beta = 0$, and the state is characterized by $T_{\hat{z}\hat{z}} = 1$ and $T_{\hat{x}\hat{x}} = \pm1$. As is well known, the positive solution identifies uniquely the state $|\Phi^+\rangle$, the negative solution the state $|\Phi^-\rangle$. These states satisfy automatically the condition (F.2). Thus,

---

[3] This paper made a few people angry for a short while. On the one side, some of those who had invested a lot of effort in proving the "unconditional security" of BB84 took some time to digest the idea that this was "unconditional under the assumption of perfectly characterized devices" (lesson to be learned: Don't take buzzwords too seriously). On the other side, the author of this book missed the opportunity to be a co-author of this breakthrough for some really silly personal issues (fortunately, the issues were recomposed and he did not miss the next one).

in order to produce the local behavior (F.1) with two qubits, one needs maximally entangled states and maximally incompatible measurements.

However, the behavior (F.1) is obviously *local*: It can be obtained by sampling, with equal probability, the four local deterministic behaviors defined as $(a_0, a_1; b_0, b_1) = (+1, +1; +1, +1)$, $(+1, -1; +1, -1)$, $(-1, +1; -1, +1)$, and $(-1, -1; -1, -1)$. Thus, the security of the BB84 protocol relies crucially on guaranteeing that signals are qubits—or more generally, that there is a "qubit fraction" in the signals [see the review (Scarani *et al.*, 2009) for details and references].

This observation led to the awareness that most tools developed in quantum information theory, from simple geometric objects like entanglement witnesses to elaborate constructions like security proofs for QKD protocols, rely on *some characterization of the devices*, usually a very accurate one. Nonlocality can dispense from this characterization. Certification through nonlocality is not just interesting for foundational concerns like guaranteeing security against hypothetical no-signaling adversaries: It is relevant for quantum information tasks as feasible today.

The next breakthrough followed from this insight: A security proof of QKD against a quantum adversary based on nonlocality (Acín, Brunner, Gisin, Massar, Pironio, and Scarani, 2007). In its title, this level of certification was called *device-independent*, and the wording stuck. This paper was the one that properly triggered the field. From a revered pioneering insight ready for storage in a museum, Bell nonlocality had become one of the most powerful tools for the certification of devices.

F.2   Overview of Tasks

Primarily, one can certify that *the device is a nonlocal resource* in a DI way. There is no exhaustive closed list of what other certifications may follow from that one. This section is a quick review of what has been considered so far.

F.2.1   Certification of secrecy tasks: Key distribution and randomness

From the history in the previous section, we know that DI started from studying *the security of key distribution* (Mayers and Yao, 1998; Barrett *et al.*, 2005a; Acín *et al.*, 2007). *A posteriori*, this was not the most logical beginning: *Randomness certification* would have been more natural, key distribution being an advanced form of randomness extraction involving two distant players. The idea of DI certification of randomness appears first in the Ph.D. thesis of Roger Colbeck under the supervision of Adrian Kent [that material was eventually published (Colbeck and Kent, 2011)]. However, the proper theoretical tools were introduced alongside with a proof-of-principle experiment[4] in (Pironio *et al.*, 2010). A further conceptual breakthrough was the introduction of randomness amplification (Colbeck and Renner, 2012), that we have presented in section 11.5.

The amount of theoretical works exploring the DI certification of randomness grew rapidly: For an overview, I refer the reader to the two review articles available on the topic (Pivoluska and Plesch, 2014; Acín and Masanes, 2016). Most works adopted the same security assumptions of QKD, with an adversary that can prepare the state and keep quantum side-information, and thus

---

[4]   That experiment used entangled ions and could certify 42 random bits. A more recent experiment with entangled photons (Shen *et al.*, 2018) certified 617,000 random bits in a 42-minute run. Both sets of authors state that the appearance of 42 was purely coincidental and is not a wink to Douglas Adams' *The Hitchhiker's Guide to the Galaxy*, where "42" is the "Answer to the Ultimate Question of Life, the Universe, and Everything".

focused on randomness expansion (which was at times termed "randomness amplification," before the latter took up the current meaning).

Experiments have already demonstrated the generation of randomness from the violation of Bell inequalities against adversaries with quantum side-information (Liu *et al.*, 2018; Shen *et al.*, 2018) or with classical side-information on no-signaling resources (Bierhorst *et al.*, 2018). An experimental demonstration of DIQKD is still lacking at the moment of writing, due to the additional requirements discussed in subsection 8.4.2.

F.2.2 Certification of quantum resources

Once one assumes that quantum theory holds, nonlocality certifies the presence of *entangled states* and the ability of performing *incompatible measurements*. As previously mentioned, one can obtain quantitative estimates, i.e., lower bounds for the *amount* of entanglement in the state, and for any measure of incompatibility of the measurements. We have not discussed these DI certifications in the main text, as the role of nonlocality is very similar to the case of randomness: Only, one minimizes a measure of entanglement rather than a measure of randomness (Moroder *et al.*, 2013). As witnesses of the versatility of DI certification, we can cite the possibility of certifying genuine multipartite entanglement (Bancal *et al.*, 2011b) and a scheme to guarantee that a measurement device has entangled eigenstates (Rabelo *et al.*, 2011).

We have devoted chapter 7 to the remarkable possibility of *self-testing*, which was discovered twice. First, the self-testing character of $S = 2\sqrt{2}$ was noticed in the context of nonlocality (Tsirel'son, 1987; Summers and Werner, 1987; Popescu and Rohrlich, 1992b). Second, by the works of Mayers and Yao mentioned—and which are so focused on cryptography that fail to mention anything about nonlocality. Strictly speaking, self-testing applies only to extremal behaviors. However, given any behavior, one can consider approximate self-testing with reference to one of these ideal cases. This makes self-testing suitable for certification. However, if a certification can be done directly, like that of randomness or entanglement, passing through an intermediate self-testing step can only worsen the bound.[5] Rather, self-testing may be helpful if the goal is to test the inner working of an all-purpose machine—a quantum computer. This line of thought was pioneered by the paper of Reichardt, Unger, and Vazirani (RUV) (2013). The main challenge of such a proof is how to go beyond the i.i.d. description: One would like to prove that observing $S \approx 2\sqrt{2}$ on N rounds certifies the presence of approximately N singlets, without even assuming a tensor product structure between the degrees of freedom measured in each round! RUV managed this theoretical feat, albeit with a very bad scaling as N grows. Further work have significantly improved the construction. For an overview of these steps and arguably the best construction at the moment of writing, we direct the reader to (Coladangelo *et al.*, 2017b). For a more pedestrian example of the self-testing of two singlets, see (Wu *et al.*, 2016).

F.2.3 Dimensionality of the physical system

The notion of randomness is independent of quantum theory, while the notion of entanglement is only defined in the theory. The notion of *dimensionality* of a physical system lies half-way. It is

---

[5] It is not obvious how to obtain a bound on, say, guessing probability from a bound on state fidelity after local isometries; but one example would suffice: Singlet fidelity $F \leq \frac{1}{2}$ are compatible with separable states, so from that bound alone one would have to infer $P_{\text{guess}} = 1$. However, if the behavior is nonlocal, there is certifiable randomness in it.

naturally defined in quantum theory as the dimensionality of the Hilbert space that describes the system. But one can give it a theory-independent meaning, as the maximal capacity of the system to encode information.

Now, one can ask: what is the *minimal dimension* of the system required to produce a given behavior? Because it depends only on the behavior, this lower bound on the dimension can be certified in a DI way. However, for this assessment nonlocality is not required (Wehner *et al.*, 2008). For instance, the local behavior (F.1) can be realized with two qubits, which is obviously the minimal dimension, but requires entanglement. It can also be realized with two four-valued strings, which have higher dimension but are trivially achievable with classical means. By rewriting results on communication complexity, one can find examples of local behaviors with an exponential gap between the classical and the quantum dimensionality, but again the latter require more and more complex entangled states. It is therefore not clear which resource is smaller, the lower-dimensional entangled state or the higher-dimensional classical realization.

When the behavior is nonlocal, no classical model can reproduce it, and a *dimension witness* based on nonlocality certifies the minimal dimension of the degrees of freedom that are entangled. The first such study proved that a sufficiently high violation of the CGLMP inequality for $m = 3$ cannot be achieved with qubits, and is therefore a witness of dimension $d \geq 3$ on both Alice and Bob (Brunner *et al.*, 2008). A short time later, it was proved that witnesses for any dimension can be found in Bell scenarios with binary outputs, increasing the number of inputs (Vértesi and Pál, 2009). More recently, it was noticed that even nonlocality-based dimension witnesses are not exempt from ambiguities (Cong *et al.*, 2017).

F.3   Device-Independent, Really?

Like every successful label, "device-independent" may lead to emotions: Those who possess it may overstate its relevance, those who don't possess it may either despise it or try to claim it for themselves. All of this has happened. Sociology of science is not the topic of this book, but a subtle confusion must be addressed. We have already hinted to it when discussing the difference between randomness and secret randomness (subsection 8.1.3).

F.3.1   Providers and adversaries

If we trust everyone's competence and honesty, there is no need for certification. Certification should be conclusive when the provider is untrusted. Misunderstanding arose from confusing the *untrusted provider* with the *adversary* as considered in cryptographic tasks.

The untrusted provider may manufacture poor quality devices: If he is aware of it, he is dishonest; if not, he is simply incompetent. The cryptographic adversary has a completely different goal: She is interested in the product of the operation of the device. To put it as a caricature: The untrusted provider may sell you an expensive handphone that does not connect to the Internet, the adversary will sell you a cheap handphone that connects perfectly well to the Internet and on which she can spy.

Let us apply this understanding to DI certification of QKD. Alice and Bob take the devices and perform a loophole-free Bell test. Based on the observed behavior, they certify that the device works as it should: It has the capacity of producing secret keys. But producing the key is not enough: It's meant to be a *secret* key, so the information should not leak out at any later time. If the device comes from the adversary, in all likelihood it contains an emitter that does exactly that: Leak out the produced keys to her at the end of the process.[6] At a price of a redundancy, let me stress that the emitter is *not* used to produce nonlocality through communication: That was duly tested by closing the locality loophole. But leakage at a later time can't be prevented based on space-like separation.

In summary, DI certification is really device-independent and applies to any device, whether manufactured by an untrusted provider (allegedly incompetent, allegedly dishonest, or both) or by the adversary. But in order to produce actual secret keys, Alice and Bob must be sure that there is no leakage. This cannot be checked in a DI way; at the very least, the adversary must not have been involved in the manufacturing.

In view of this discussion, some DI studies have assumed that the devices are manufactured by *honest*[7] *providers* (e.g., our colleagues in academia). One may then accept to relax some of the requirements for a loophole-free Bell test. Notably, notice how to activate the locality loophole one must maliciously engineer communication channels that specifically send information about the inputs, whereas the detection loophole could be opened unwittingly (recall Appendix B.3). Thus, an honest provider may be dispensed from closing the locality loophole, while the detection loophole must be closed to claim DI.

F.3.2   Of labels and men

If a certification leaves some loopholes open because this is reasonable under some model of provider, does it still deserve the label "device-independent"? Is it accurate to speak of "device-independent security of QKD," given that no leakage must be assumed to actually produce secret keys? I prefer not to take strong stances on these matters. Ultimately, labels have a limited value for scientists.

What matters for this book is that we appreciate the unique power of nonlocality as a tool of certification. What matters for science in general is that *the assumptions underlying the claims are clearly stated.* In this last respect, the field of DI has made a positive impact on the whole of quantum information. In particular, there has been a growing interest in intermediate levels of certification. We devote section F.4 to them.

F.4   Certification with Partially Characterized Devices

Inspired by DI certification, several certifications have been proposed that rely on a *partial characterization of the devices.* The following list cannot be exhaustive: Anyone who has a good reason to make an assumption should feel free to define the corresponding certification. It is always assumed that quantum theory is valid.

F.4.1   Characterizing specific devices

This form of departure from DI certification is of particular appeal to experimentalists who (justifiably) are pretty confident of their understanding of the working of their devices.

---

[6] A manufacturing adversary may use subtler, more complicated tricks to learn the key in the context of QKD (Barrett *et al.*, 2013).

[7] Usually termed *trusted* providers in the literature, but we saw that the wording is ambiguous.

- The *assumption of fair-sampling at detection* is the most famous such example (recall subsection 1.5.2). The attitude of the community towards this assumption has undergone fluctuations. For several years, it had been considered absolutely natural. Experimental groups were working towards closing the detection loophole for the glory of having done it and for the exciting technical challenges that it entailed, but nobody was taking the threat seriously. The mood swung almost completely at the height of excitement for DI certification, when theorists started pushing for taking the detection loophole as a serious threat—and they were listened to, although an actual experiment played also an important role (Gerhardt *et al.*, 2011).

  That being said, it would be too sectarian to put an anathema on every experiment that does not close the detection loophole: Strong as it may be, fair-sampling remains a legitimate assumption to be stated, especially if one is trying to demonstrate something other than strict Bell nonlocality or DI certification.Besides, the fair-sampling assumption can be relaxed to a Santha-Vazirani assumption by adapting the certification of nonlocality under measurement dependence (chapter 11). With these techniques, it is enough to assume that, in each round, each detector must have non-zero probability of firing (Pütz *et al.*, 2016).

- One could assume that a device other than the detector behaves "as it should." For instance, by assuming that a beam-splitter does not introduce any temporal delay, one can demonstrate nonlocality in experiments with time-bin entanglement that otherwise have a local variable description (Martin *et al.*, 2013).

- Bell nonlocality with optical fields is a sub-field that we have not discussed in this book. While there exist proposals of proper Bell tests (Banaszek and Wódkiewicz, 1998; García-Patrón *et al.*, 2004; Żukowski *et al.*, 2016), a lot of works in the literature relied on optical assumptions. This is due to the fact that it's easy to find Bell inequalities for measurements of the photon number N, but no easily-implemented measurement gives access to this degree of freedom, while the X and P quadrature operators are associated to the commonplace homodyne or heterodyne measurements. Now, in quantum optics, it holds $N = \frac{1}{2}(X^2 + P^2 - 1)$ for a monomode field. Thus, one can perform quadrature measurements, derive a value of N in each round, and compute its statistics. However, if one were to replace N with X and P into the Bell inequality, the resulting expression can be violated with LVs.

- The same kind of assumption is made in the verification of a family of Bell inequalities for many-body systems (Tura *et al.*, 2014; Schmied *et al.*, 2016). Once again, the theoretical objects are proper Bell inequalities that contain only two-body correlators. Two-body correlators can be measured in atomic gases, but only with all particles subject to the same input; while the inequality needs also correlators when the two inputs are different. Adding the notion that the measurement of a spin is a measurement of a direction, the desired correlators can be extracted by comparing feasible measurements along different directions. Indeed, $\vec{n} \cdot \vec{m} = \frac{1}{2}[|\vec{n} + \vec{m}|^2 - |\vec{n}|^2 - |\vec{m}|^2]$.

- Finally, some certifications have been made under the assumption that one knows the *dimension of the Hilbert space* of the system under study, but nothing about the measurements. In abstract terms, the assumption is valid; but it is not clear on what evidence it can be based. Recently, a much more promising approach has been proposed to deal with more physical assumptions like upper bounds on the energy or on the number of photons (Van Himbeeck *et al.*, 2017).

F.4.2   Steering: Uncharacterized Alice, characterized Bob

Many presentations of entanglement stress the idea that, by making a measurement, Alice can *steer* Bob's state into one or another basis. For instance, if the shared state is $|\Phi^+\rangle_{AB}$, by measuring her qubit in the Z_A basis, Alice will prepare either $|+\hat{z}\rangle_B$ or $|-\hat{z}\rangle_B$ on Bob's side. But she could have measured in the X_A basis, and then prepare either $|+\hat{x}\rangle_B$ or $|-\hat{x}\rangle_B$. So, Alice could tell Bob in each round which state he is supposed to have. This looks impressive because we have fully characterized the joint state of both Alice and Bob and we know it to be $|\Phi^+\rangle_{AB}$. It would look far less impressive if that state would have been an incoherent mixture of $|0\rangle_A |+\hat{z}\rangle_B$, $|1\rangle_A |-\hat{z}\rangle_B$, $|2\rangle_A |+\hat{x}\rangle_B$, and $|3\rangle_A |-\hat{x}\rangle_B$: In this case, Alice's measurement in the computational basis reveals which state already exists on Bob's side, so there is no steering. Could Bob be convinced that Alice can steer his state without knowing anything about Alice's degree of freedom?

This question led to the formal characterization of *EPR-steering* (Wiseman *et al.*, 2007): Alice won't be able to convince Bob that she is steering his state if there exist normalized states ρ_λ of Bob's system such that

$$\rho_B = \int d\lambda q(\lambda)\rho_\lambda \tag{F.4}$$

and the observed behavior can be written

$$P(a, b|x, y) = \int d\lambda q(\lambda)P_\lambda(a|x)\mathrm{Tr}(\rho_\lambda \Pi_b^y). \tag{F.5}$$

It is evident that Bell nonlocality implies steering, as (F.5) is a local behavior. Also, if the joint state ρ_{AB} is separable, obviously (F.5) holds together with (F.4). That being said, EPR-steering is provably different from both nonlocality and tomography: There exist local behavior that demonstrate steering, and there exist entangled states whose state-behavior cannot demonstrate steering. For many more results and all references, we refer to reader to the review (Cavalcanti and Skrzypczyk, 2017).

F.4.3   Characterized quantum inputs

Francesco Buscemi (2012) introduced the idea of Bell-type tests in which the inputs are quantum states: To any $x \in \mathcal{X}$ there corresponds a state ξ_x, and to any $y \in \mathcal{Y}$ there corresponds a state υ_y. The players know the possible states but may not know which state is sent in each round. He proved several results in this scenario.[8] We are interested in a corollary: Given *any* shared entangled state ρ_{AB}, in this scenario the players are able to verify that the state is entangled indeed. He proved that every entangled state can be detected in such a scenario. The connection with certification was first emphasized in (Cavalcanti *et al.*, 2013). One can turn these techniques into a fully DI detection of entanglement by self-testing the input states (Bowles *et al.*, 2018).

---

[8]   Notably, he could define a necessary and sufficient condition for the possibility of transforming one state into another with local operations and shared randomness (LOSR).

Following (Branciard *et al.*, 2013), let us first describe the procedure with fully characterized devices. As mentioned in Appendix C.2, an entanglement witness is an operator W such that $\mathrm{Tr}(\rho W) \geq 0$ for all separable states ρ and $\mathrm{Tr}(\rho W) < 0$ for some entangled states. An entanglement witness for two-qudit states can always be decomposed, in a non-unique way, as

$$W = \sum_{x,y} w_{xy} \xi_x^T \otimes \upsilon_y^T \tag{F.6}$$

where the ξ_x and υ_y are qudit states, T indicates transpose and w_{xy} are some real coefficients. Suppose Alice and Bob share the two-qudit state ρ_{AB}. If Alice is given an auxiliary system in the state ξ_x, she can perform on it and her system the joint projective measurement that yields $a = 1$ for the state $\frac{1}{\sqrt{d}} \sum_i |i\rangle \otimes |i\rangle$ and zero otherwise; Bob follows the analog measurement scheme. It is then easy to prove that $P(1,1|x,y) = \mathrm{Tr}[\rho_{AB}(\xi_x^T \otimes \upsilon_y^T)]$, so that $\sum_{x,y} w_{xy} P(1,1|x,y) = \mathrm{Tr}(\rho_{AB} W)$. By definition of an entanglement witness,

$$\sum_{x,y} w_{xy} P(1,1|x,y) < 0 \tag{F.7}$$

is only possible if ρ_{AB} is entangled. Conversely, since entanglement witnesses exist for any entangled ρ_{AB}, every entangled state can be detected by a suitable choice of the input states ξ_x and υ_y.

The narrative of certification with partial characterization follows straightforwardly: The players claim to have the state ρ_{AB} that is entangled, and the verifier wants to check that. In each round, the verifier sends one of the states ξ_x to Alice and one of the states υ_y to Bob. The players return their binary outputs $a, b \in \{0, 1\}$, and the verifier can compute (F.7).

F.4.4   Networks with uncorrelated sources (a.k.a. *N*-locality)

The setting for this last subsection is that of a network configuration. To fix the idea, consider first *entanglement swapping*. In a fully characterized description, Alice prepares the entangled pair and sends one of the subsystems to Charlie; so does Bob; then Charlie performs an entangling measurement, conditioned on whose outcome the systems of Alice and Bob end up entangled. In a fully device-independent description, Alice, Bob, and Charlie are the players and may have prepared any joint strategy: What happens in the intermediate stages stays in the black box, and the whole procedure is interpreted as a source creating entanglement between Alice and Bob.

But from a physics perspective, the physicists operate devices that are source of entanglement: It is very natural to assume that Alice's and Bob's sources are *uncorrelated* (Figure F.1, top). Under this assumption, called *bilocality*, local three-partite behaviors are described by

$$P(a, b, c|x, y, z) = \int d\lambda_A \int d\lambda_B q_A(\lambda_A) q_B(\lambda_B) P(a|x, \lambda_A) P(b|y, \lambda_B) P(c|z, \lambda_A, \lambda_B) \tag{F.8}$$

which is more restrictive than the general $\int d\lambda q(\lambda) P(a|x,\lambda) P(b|y,\lambda) P(c|z,\lambda)$. It should be clear how to extend the idea to other networks, and one would speak of *N-locality* or *network-locality*.

The characterization of *N*-locality is challenging, notably because the set of *N*-local behaviors is not convex. There is no comprehensive review at the moment of writing, but the reader can gather

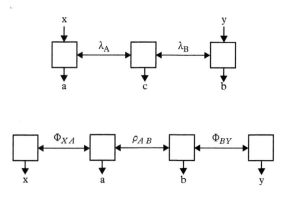

Figure F.1 *Top: Bilocality scenario modeling the physical understanding of entanglement swapping: Nonlocality is probed between Alice and Bob; the resources λ_A and λ_B are independent, so all correlation must come from conditioning on the output c. Bottom: A three-locality scenario with no inputs that describes a normal Bell test on A and B, with measurement independence explicitly enforced.*

a good amount of information from (Branciard *et al.*, 2012b,a; Fraser and Wolfe, 2018; Renou *et al.*, 2019).

Let me finish with an intriguing remark. Consider the three-locality scenario of Figure F.1, bottom. It has no inputs. The shared resources Φ_{XA} and Φ_{BY} are such that both players obtain the same output in each round (you can think of a maximally entangled state measured in the computational basis, but it could be the corresponding local resource too). Thus, Alice can use her half of $|\Phi\rangle_{XA}$ to generate the input x for a Bell test on ρ_{AB}, and Bob can behave similarly. In this sense, Figure F.1 bottom could be read as a version of Figure 1.1, describing a generic Bell test: Indeed, the maximally entangled states are arguably the most perfect realization of a quantum random number generator that reveals its output, and the assumption of three-locality is nothing else than measurement independence for the Bell test to be performed on ρ_{AB}.

Appendix G
Repository of Technicalities

This appendix contains some instructive technical results that would have slowed down the pace of the main text.

G.1 Analytical Solution of the Facets of the CHSH Correlation Polytope

It is instructive to work out explicitly the facets of $\mathcal{L}$ and derive the corresponding Bell inequalities. The simplest scenario has $M_A = M_B = m_A = m_B = 2$. In this case, $D_{NS} = 8$ and there are 16 extremal points: Finding the facets is a very easy task for a computer, but still cumbersome to write down here. We are rather going to study a very meaningful sub-polytope of $\mathcal{L}$.

For a choice of settings (x, y) and binary outputs, the *correlation coefficient* is defined by

$$E_{xy} = P(a = b|x,y) - P(a \neq b|x,y). \tag{G.1}$$

Any quadruple of numbers

$$u = (E_{00}, E_{01}, E_{10}, E_{11}) \tag{G.2}$$

with $-1 \leq E_{xy} \leq 1$ is *a priori* a valid correlation vector. The sixteen vectors such that $||u||^2 = 4$, i.e., those vectors whose components are either $+1$ or -1 are extremal points of a polytope embedded in $\mathbb{R}^4$.

To see which constraints are added by requiring that $\mathcal{P} \in \mathcal{L}$, it is convenient to use the labeling convention $a, b \in \{-1, +1\}$. With this choice, $E_{xy} = \langle a_x b_y \rangle$. In particular, for LD processes it holds $E_{xy} \overset{LD}{=} a_x b_y$, which directly leads to $E_{00} E_{01} E_{10} E_{11} \overset{LD}{=} 1$. Therefore, the extremal points of the *local correlation polytope* are the eight vectors

$$
\begin{aligned}
v_1 &= (+1, +1, +1, +1), & v_5 &= -v_1 = (-1, -1, -1, -1) \\
v_2 &= (+1, +1, -1, -1), & v_6 &= -v_2 = (-1, -1, +1, +1) \\
v_3 &= (+1, -1, +1, -1), & v_7 &= -v_3 = (-1, +1, -1, +1) \\
v_4 &= (+1, -1, -1, +1), & v_8 &= -v_4 = (-1, +1, +1, -1).
\end{aligned}
\tag{G.3}
$$

Notice that $\{v_1, v_2, v_3, v_4\}$ are mutually orthogonal, so in particular they are linearly independent: This implies that $\mathbb{R}^4$ is the smallest embedding for the local correlation polytope. Now we want to characterize the facets of this polytope. Four linearly independent vectors are required to define a

3-dimensional hyperplane:[1] Our task consists of listing all sets of four linearly independent extremal points, constructing the hyperplane that they generate, and checking if it is indeed a facet.

The symmetry of the problem makes the task simple. There are sixteen sets of four linearly independent vectors, namely the $V_{\underline{s}} = \{s_1 v_1, s_2 v_2, s_3 v_3, s_4 v_4\}$ with $\underline{s} = [s_1, s_2, s_3, s_4] \in \{-1, +1\}^4$. The normal to the hyperplane generated by $V_{\underline{s}}$ is the solution to the four linearly independent equation $n_{\underline{s}} \cdot (s_k v_k) = f_{\underline{s}} \equiv 4$ for $k = 1, \ldots, 4$ (the constant being chosen for simplicity). It is readily found to be $n_{\underline{s}} = \sum_{k=1}^4 s_k v_k$, either by direct inspection or by noticing that $v_i \cdot v_j = 4\delta_{ij}$ for $1 \le i, j \le 4$. Moreover, the extremal points that do not sit on the hyperplane defined by $V_{\underline{s}}$ are the four $-s_k v_k$, which all lie on the same side of the hyperplane since obviously $n_{\underline{s}} \cdot (-s_k v_k) = -4$. Therefore each of the sixteen sets $V_{\underline{s}}$ defines a facet by the condition

$$n_{\underline{s}} \cdot u \le 4 \text{ with } n_{\underline{s}} = \sum_{k=1}^4 s_k v_k. \tag{G.4}$$

To finish our study, we just have to inspect each of these facets (since $n_{-\underline{s}} = -n_{\underline{s}}$, we can just look at eight of them). We find:

$$
\begin{aligned}
n_{[+1,+1,+1,+1]} &= (4, 0, 0, 0) &\longrightarrow&\quad 4E_{00} \le 4, \\
n_{[+1,+1,+1,-1]} &= (2, 2, 2, -2) &\longrightarrow&\quad 2E_{00} + 2E_{01} + 2E_{10} - 2E_{11} \le 4, \\
n_{[+1,+1,-1,+1]} &= (2, 2, -2, 2) &\longrightarrow&\quad 2E_{00} + 2E_{01} - 2E_{10} + 2E_{11} \le 4, \\
n_{[+1,-1,+1,+1]} &= (2, -2, 2, 2) &\longrightarrow&\quad 2E_{00} - 2E_{01} + 2E_{10} + 2E_{11} \le 4, \\
n_{[+1,+1,-1,-1]} &= (0, 4, 0, 0) &\longrightarrow&\quad 4E_{01} \le 4, \\
n_{[+1,-1,+1,-1]} &= (0, 0, 4, 0) &\longrightarrow&\quad 4E_{10} \le 4, \\
n_{[+1,-1,-1,+1]} &= (0, 0, 0, 4) &\longrightarrow&\quad 4E_{11} \le 4, \\
n_{[+1,-1,+1,+1]} &= (-2, 2, 2, 2) &\longrightarrow&\quad -2E_{00} + 2E_{01} + 2E_{10} + 2E_{11} \le 4.
\end{aligned} \tag{G.5}
$$

In summary, the local correlation polytope for the simplest scenario has sixteen facets. Up to relabeling of the inputs and/or of the outputs, eight of these describe the constraint $E_{00} \le 1$ and are therefore trivial, while the other eight describe the constraint

$$S \equiv E_{00} + E_{01} + E_{10} - E_{11} \le 2. \tag{G.6}$$

This constraint is not trivial, and indeed can be violated by valid correlation vectors which do not belong to the local polytope: In particular, the vector $w = (+1, +1, +1, -1)$ reaches up to $S = 4$. In fact, we know that this is the CHSH inequality (2.29).

G.2 Hardy's Test from the Schmidt Decomposition of the State

This appendix repeats the calculation of Hardy's test (subsection 4.4.1) starting from the two-qubit pure state written in the Schmidt decomposition (3.15).

---

[1] This is because the origin $(0, 0, 0, 0)$ is inside the polytope. If the origin were on a facet, only three linearly independent vectors would be sufficient to specify such a facet. This is a technical point that does not need to bother us here, but may play a role for most compact parametrizations of probability polytopes.

Without loss of generality, the direction of Alice's first measurement can be written

$$\hat{a}_0 = \cos\alpha_0 \hat{z} + \sin\alpha_0 \hat{x}. \tag{G.7}$$

This choice, together with the constraints (1.11), will fix all the other measurements in a sequential way. It is convenient to work with the projector on the state (3.15)

$$\Pi_\theta = \frac{1}{4}\left[\mathbb{I}\otimes\mathbb{I} + \cos 2\theta\,(\sigma_{\hat{z}}\otimes\mathbb{I} + \mathbb{I}\otimes\sigma_{\hat{z}}) + \sigma_{\hat{z}}\otimes\sigma_{\hat{z}} + \sin 2\theta\,(\sigma_{\hat{x}}\otimes\sigma_{\hat{x}} - \sigma_{\hat{y}}\otimes\sigma_{\hat{y}})\right].$$

The output $+1$ of Alice's measurement is associated with the projector $\Pi_{A_0}^+ = \frac{1}{2}(\mathbb{I} + \hat{a}_0\cdot\sigma)$; when this output is found, the state prepared on Bob's side must correspond to $\Pi_{B_0}^-$ in order to satisfy the constraint $P(a_0 = b_0 = +1) = 0$. Noting that $\Pi_{A_0}^+\sigma_{\hat{z}}\Pi_{A_0}^+ = \cos\alpha_0\,\Pi_{A_0}^+$, $\Pi_{A_0}^+\sigma_{\hat{x}}\Pi_{A_0}^+ = \sin\alpha_0\,\Pi_{A_0}^+$, and $\Pi_{A_0}^+\sigma_{\hat{y}}\Pi_{A_0}^+ = 0$, one finds $\Pi_{A_0}^+\Pi_\theta\Pi_{A_0}^+ = \frac{1+\cos 2\theta\cos\alpha_0}{2}\Pi_{A_0}^+\otimes\Pi_{B_0}^-$ with

$$\Pi_{B_0}^- = \frac{1}{2}\left(\mathbb{I} + \frac{\cos 2\theta + \cos\alpha_0}{1 + \cos 2\theta\cos\alpha_0}\sigma_{\hat{z}} + \frac{\sin 2\theta\sin\alpha_0}{1 + \cos 2\theta\cos\alpha_0}\sigma_{\hat{x}}\right).$$

That is, we have found

$$\hat{b}_0 = \cos\beta_0\hat{z} + \sin\beta_0\hat{x},\ \text{with}\ \begin{cases}\cos\beta_0 = -\frac{\cos 2\theta + \cos\alpha_0}{1 + \cos 2\theta\cos\alpha_0}\\ \sin\beta_0 = -\frac{\sin 2\theta\sin\alpha_0}{1 + \cos 2\theta\cos\alpha_0}\end{cases}. \tag{G.8}$$

The calculation for $\Pi_{A_0}^-$ follows by replacing α_0 with $\pi + \alpha_0$, and because of the second constraint we must read it as $\Pi_{A_0}^-\Pi_\theta\Pi_{A_0}^- = \frac{1-\cos 2\theta\cos\alpha_0}{2}\Pi_{A_0}^-\otimes\Pi_{B_1}^-$. Therefore

$$\hat{b}_1 = \cos\beta_1\hat{z} + \sin\beta_1\hat{x},\ \text{with}\ \begin{cases}\cos\beta_1 = -\frac{\cos 2\theta - \cos\alpha_0}{1 - \cos 2\theta\cos\alpha_0}\\ \sin\beta_1 = \frac{\sin 2\theta\sin\alpha_0}{1 - \cos 2\theta\cos\alpha_0}\end{cases}. \tag{G.9}$$

Finally, the third constraint leads to $\Pi_{B_0}^-\Pi_\theta\Pi_{B_0}^- = \frac{1-\cos 2\theta\cos\beta_0}{2}\Pi_{A_1}^-\otimes\Pi_{B_0}^-$ with[2]

$$\hat{a}_1 = \cos\alpha_1\hat{z} + \sin\alpha_1\hat{x},\ \text{with}\ \begin{cases}\cos\alpha_1 = -\frac{2\cos 2\theta + \cos\alpha_0(1+\cos^2 2\theta)}{1 + \cos^2 2\theta + 2\cos 2\theta\cos\alpha_0}\\ \sin\alpha_1 = -\frac{\sin^2 2\theta\sin\alpha_0}{1 + \cos^2 2\theta + 2\cos 2\theta\cos\alpha_0}\end{cases}. \tag{G.10}$$

Using (G.9) and (G.10), after some tedious algebra one gets

$$P(+, +|1, 1) = \frac{1}{4}\left(1 + \cos 2\theta\,(\cos\alpha_1 + \cos\beta_1) + \cos\alpha_1\cos\beta_1 + \sin 2\theta\sin\alpha_1\sin\beta_1\right)$$

$$= \frac{1}{2}\frac{\sin^2\alpha_0\cos^2 2\theta\sin^2 2\theta}{(1 + \cos^2 2\theta + 2\cos 2\theta\cos\alpha_0)(1 - \cos 2\theta\cos\alpha_0)}. \tag{G.11}$$

---

[2] The coefficients in (G.10) are obtained by taking (G.9), replacing $\beta_1 \to \alpha_1$ and $\alpha_0 \to \beta_0$, then plugging in (G.8).

The maximization over the choice of measurement direction yields

$$\max_{\alpha_0} P(+, +|1, 1) = \frac{\sin^2 2\theta(1 - \sin 2\theta)}{(2 - \sin 2\theta)^2} \quad \text{for} \quad \cos \alpha_0 = -\frac{1 - \sin 2\theta}{\cos 2\theta}. \tag{G.12}$$

We find $P(+, +|1, 1) = 0$ only for $\theta = 0$ and $\theta = \frac{\pi}{4}$. The first condition identifies product states, which are expected not to show any nonlocality. The second identifies maximally entangled states, for which $\hat{a}_0 = -\hat{a}_1 = -\hat{b}_0 = \hat{b}_1$: Both on Alice and on Bob, the measurements are identical, so nonlocality can't be demonstrated. For every other entangled pure state of two qubits, $0 < \theta < \frac{\pi}{4}$, $P(+, +|1, 1)$ is strictly positive for every choice of A_0 other than $\sigma_{\hat{z}}$ the local Schmidt basis. Finally, the maximal quantum violation is indeed (4.17) obtained for the state with $\sin 2\theta = 3 - \sqrt{5}$.

G.3   Pitfalls in Handling Signaling Behaviors

This appendix is devoted to warning against possible pitfalls when dealing with signaling behaviors in a too abstract way. The safest way to avoid such pitfalls is to describe the signaling resource, in particular the directionality of the signaling.

We work in the $(2, 2; 2, 2)$ Bell scenario and focus on the signaling behavior defined by

$$P(0, 0|x, 0) = P(1, 0|x, 1) = 1 \text{ for } x \in \{0, 1\}, \tag{G.13}$$

all the other probabilities being zero by normalization. It is easy to express in words what this behavior describes: Alice's input x is ignored, Alice's output a is equal to Bob's input y (this is the signaling from Bob to Alice), and Bob's output b is deterministically 0.

Clearly, no resource that produces this behavior can output a before y has been provided by Bob. Such timing issue is not a concern for no-signaling behaviors, with which researchers of nonlocality are familar. In particular, as mentioned in section 5.3, the Svetlichny polytope does not take it into consideration. This is why the warning must be raised (Gallego *et al.*, 2012).

To see the dangers, let us work *ex absurdo* and suppose that, somehow, there exists a resource that produces the behavior (G.13) irrespective of the time ordering. Suppose then that Alice inputs x and immediately sees an output a: This means that Bob will be obliged to input $y = a$, which is absurd if Bob is at a distance. This intuitive absurdity can be even translated into mathematical absurdity by supposing that Alice, through another channel, sends a to Bob. If Bob inputs a, the probability of him observing 0 would be computed as

$$P(b = 0|x) = \sum_a P(a|x)P(b = 0|y = a; a, x) = \sum_a P(a, b = 0|x, y = a) = 2.$$

Similarly, if Bob inputs $y = 1 - a$, one would compute

$$P(b = 0|x) = \sum_a P(a|x)P(b = 0|y = 1 - a; a, x) = \sum_a P(a, b = 0|x, y = 1 - a) = 0$$

which is also absurd given that, by inspection of the behavior, b is never equal to 1 either. The formal mistake in both equations consists in writing $P(a, b|x, y) = P(a|x)P(b|y; a, x)$ instead of $P(a, b|x, y) = P(a|x, \underline{y})P(b|y; a, x)$, matching the absurd assumption we started from.

As noticed by (Gallego *et al.*, 2012), the careless usage of this model in the Svetlichny context may lead to predict nonlocality where there can't possibly be any. Indeed, it is easy to predict a high value of CHSH if one's probabilities sum up to 2...

G.4 Jordan's Lemma

Jordan's lemma states the following. Let A_0 and A_1 be Hermitian operators on an arbitrary Hilbert space $\mathcal{H}$, with eigenvalues -1 and $+1$. There exists a basis in which both operators are block-diagonal, in blocks of dimension 2×2 at most.

G.4.1 Proof of Jordan's lemma

By definition, $A_0^2 = A_1^2 = \mathbb{I}_{d_A}$. Besides, $U = A_0 A_1$ is unitary. Indeed, let us denote $|\alpha, 0\rangle$ an eigenstate of U: $U |\alpha, 0\rangle = \omega_\alpha |\alpha, 0\rangle$ with $|\omega_\alpha| = 1$. Then $|\alpha, 1\rangle = A_0 |\alpha, 0\rangle$ is also an eigenstate of U, since $U |\alpha, 1\rangle = A_0 A_1 A_0 |\alpha, 0\rangle = A_0 U^\dagger |\alpha, 0\rangle = \omega_\alpha^* |\alpha, 1\rangle$. Therefore:

$$A_0 |\alpha, 0\rangle = |\alpha, 1\rangle \text{ and } A_0 |\alpha, 1\rangle = |\alpha, 0\rangle, \tag{G.14}$$

$$A_1 |\alpha, 1\rangle = U^\dagger |\alpha, 0\rangle = \omega_\alpha^* |\alpha, 0\rangle \text{ and } A_1 |\alpha, 0\rangle = A_1 (\omega_\alpha A_1 |\alpha, 1\rangle) = \omega_\alpha |\alpha, 1\rangle. \tag{G.15}$$

At this point, there are two possibilities:

- $|\alpha, 0\rangle \equiv |\xi\rangle$ is an eigenvector of A_0 for the eigenvalue $\lambda_{\xi, 0} \in \{-1, +1\}$. It follows from what precedes that $|\alpha, 1\rangle = \lambda_{\xi, 0} |\xi\rangle$, and that $|\xi\rangle$ is also an eigenvector of A_1 for the eigenvalue $\lambda_{\xi, 1} = \omega_\xi \lambda_{\xi, 0}$ (whence $\omega_\xi = -1$ or $+1$).
- $|\alpha, 0\rangle$ is not an eigenvector of A_0. In this case, $\langle \alpha, 0 | \langle \alpha, 0 \rangle = 0$, since they are different eigenvectors of the normal operator U. Now the conditions (G.14) read $A_0|_\alpha = \sigma_{\hat{x}}^\alpha$ and the conditions (G.15) read $A_1|_\alpha = \text{Re}(\omega_\alpha)\sigma_{\hat{x}}^\alpha - \text{Im}(\omega_\alpha)\sigma_{\hat{y}}^\alpha$ where the σ_k^α are the usual Pauli matrices defined in $\text{Span}\{|\alpha, 0\rangle, |\alpha, 1\rangle\}$ with the convention $\sigma_{\hat{z}}^\alpha = |\alpha, 0\rangle \langle \alpha, 0| - |\alpha, 1\rangle \langle \alpha, 1|$.

Since the eigenvectors of a unitary operator span the whole Hilbert space, we have proved that

$$A_0 = \bigoplus_\alpha \sigma_{\hat{x}}^\alpha \bigoplus_\xi \lambda_{\xi, 0} |\xi\rangle \langle \xi|, \tag{G.16}$$

$$A_1 = \bigoplus_\alpha \left[\text{Re}(\omega_\alpha)\sigma_{\hat{x}}^\alpha - \text{Im}(\omega_\alpha)\sigma_{\hat{y}}^\alpha \right] \bigoplus_\xi \lambda_{\xi, 1} |\xi\rangle \langle \xi|. \tag{G.17}$$

This concludes the proof.

G.4.2 Usefulness of the lemma

Jordan's lemma has been used often in the early days of device-independent certification, as a tool to reduce calculations from unknown Hilbert spaces to qubits. Unfortunately, it cannot be extended to more operators and/or more eigenvalues, hence its rather limited value.

In the main text, we invoked Jordan's lemma in the proof of self-testing of $S = 2\sqrt{2}$. Specifically, in subsection 7.1.1 we claimed that: If there is a subspace of $\{A_0, A_1\}$ associated to the eigenvalue 0, it must be of even dimension $2d''_A$, and that there exist a choice of basis in which $A_0 = \sigma_{\hat{z}} \otimes \mathbb{I}_{d''_A}$ and $A_1 = \sigma_{\hat{x}} \otimes \mathbb{I}_{d''_A}$. This is indeed a corollary of Jordan's lemma: In the subspaces indexed by β, $\{A_0, A_1\}|\beta\rangle = 2\omega_\beta \in \{-2, +2\}$, so to find the eigenvalue 0 we need to stay in the subspaces indexed by α, which are even-dimensional. Furthermore,

$$\{\sigma_{\hat{x}}^\alpha, \text{Re}(\omega_\alpha)\sigma_{\hat{x}}^\alpha - \text{Im}(\omega_\alpha)\sigma_{\hat{y}}^\alpha\} = \text{Re}(\omega_\alpha)\mathbb{I}$$

because $\{\sigma_{\hat{x}}^\alpha, \sigma_{\hat{x}}^\alpha\} = 0$. Therefore to have the eigenvalue 0 one has to restrict to sectors such that $A_1|_\alpha = \sigma_{\hat{y}}^\alpha$. The claim made in the text follows by choosing a basis in which $\sigma_{\hat{x}}^\alpha \equiv \sigma_{\hat{z}}$ and $\sigma_{\hat{y}}^\alpha \equiv \sigma_{\hat{x}}$ for all α under consideration.

G.5   A Case Study of Randomness with Characterised Devices

G.5.1   Introduction and basic result

We consider the example of extraction of randomness from characterized devices discussed in 8.2.4: Claire sees $\rho = \frac{1}{2}(\mathbb{I} + \eta\sigma_{\hat{z}})$ and performs the projective measurement of $\sigma_{\hat{x}}$. Eve's most favorable decomposition of ρ is easy to guess (Figure G.1), and optimality can indeed be proved easily assuming that the device is really producing two pure states with equal probability, such that $\rho = \rho_+ + \rho_-$ with $\rho_\pm = \frac{1}{2}|\pm\chi\rangle\langle\pm\chi|$. The proof of optimality starting without any assumption is cumbersome and we leave it aside.

In summary, we have to assume that Eve sees one of the two states

$$|\chi_\pm\rangle = \sqrt{\frac{1}{2}(1 \pm \sqrt{1 - \eta^2})}|+\hat{x}\rangle + \sqrt{\frac{1}{2}(1 \mp \sqrt{1 - \eta^2})}|-\hat{x}\rangle. \tag{G.18}$$

Upon seeing $|\chi_\gamma\rangle$, she guesses $c = \gamma$ and the guess is correct with probability

$$P_g = \frac{1}{2}(1 + \sqrt{1 - \eta^2}). \tag{G.19}$$

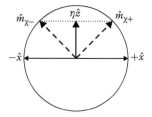

Figure G.1 *Bloch sphere representation of the randomness scenario of appendix G.5, with the decomposition of Claire's mixed state that optimizes Eve's probability of guessing the output bit.*

Notice in particular that

$$\langle +\hat{x}|\chi_+\rangle = \langle -\hat{x}|\chi_-\rangle = \sqrt{P_g}, \quad \langle +\hat{x}|\chi_-\rangle = \langle -\hat{x}|\chi_+\rangle = \sqrt{1 - P_g} \tag{G.20}$$

which will be useful later.

Now we are going to use this toy model to exhibit a difference between classical and quantum side-information. To do so, we consider *two rounds of the process, at the end of which the parity of the two bits is revealed*. We are going to compute what is the probability that Eve guesses correctly the first bit (the other being then trivially inferred).

G.5.2   Calculation with classical side-information

Due to the symmetry of the problem, we can assume that Eve sees the states $|\chi_+\rangle|\chi_+\rangle$ in the two rounds. Claire's bit is then distributed according to the following $P_{c_1 c_2}$: $P_{++} = P_g^2, P_{+-} = P_{-+} = P_g(1 - P_g), P_{--} = (1-P_g)^2$. Without further information, she would guess that the first bit is $+1$ with probability P_g.

Now Claire reveals the parity:

- In a fraction $P_{\text{even}} = P_g^2 + (1 - P_g)^2$ of the rounds, Claire will announce $c_1 c_2 = 1$. Then Eve's guess for the first bit remains $+1$, correct with conditional probability $\frac{P_g^2}{P_{\text{even}}}$.
- In a fraction $P_{\text{odd}} = 2P_g(1 - P_g)$ of the rounds, Claire will announce $c_1 c_2 = -1$. In these rounds, Eve's guess is random: Let's say that she sticks to $+1$, but she'll be correct with conditional probability $\frac{1}{2}$.

On average over all the rounds, Eve's probability of guessing the first bit is

$$P_g' = P_{\text{even}} \times \frac{P_g^2}{P_{\text{even}}} + P_{\text{odd}} \times \frac{1}{2} = P_g. \tag{G.21}$$

In other words, the guessing probability is unchanged! The fact that Claire revealed the parity helps only insofar as, if Eve correctly guesses the first bit, she will correctly guess the second as well.

G.5.3   Calculation with quantum side-information

In this case, Eve does not see a particular pair of states in each round: Rather, Claire's perceived mixture comes about because Eve is entangled with her system and keeps the purification. So, for our two-round experiment, the Claire-Eve state is

$$|\Psi\rangle_{AE} = \left[\frac{1}{\sqrt{2}} \left(|\chi_+\rangle |e_1\rangle + |\chi_-\rangle \left| e_1^\perp \right\rangle \right) \right] \otimes \left[\frac{1}{\sqrt{2}} \left(|\chi_+\rangle |e_2\rangle + |\chi_-\rangle \left| e_2^\perp \right\rangle \right) \right]. \tag{G.22}$$

Now Claire reveals the parity:

- With probability $\frac{1}{2}$ she announces $c_1 c_2 = 1$. Then Eve knows that she has one of the two states

$$\psi_{E,++} = \left(\sqrt{P_g}\,|e_1\rangle + \sqrt{1-P_g}\,\Big|e_1^\perp\Big\rangle\right) \otimes \left(\sqrt{P_g}\,|e_2\rangle + \sqrt{1-P_g}\,\Big|e_2^\perp\Big\rangle\right)$$
$$\psi_{E,--} = \left(\sqrt{1-P_g}\,|e_1\rangle + \sqrt{P_g}\,\Big|e_1^\perp\Big\rangle\right) \otimes \left(\sqrt{1-P_g}\,|e_2\rangle + \sqrt{P_g}\,\Big|e_2^\perp\Big\rangle\right)$$

with equal *a priori* probability; we have used (G.20). She would like to find out which one she actually has. The classic result by Hesltröm says that the highest probability of correct guessing in such a task is given by $\frac{1}{2}(1+\sqrt{1-|\langle\psi_{E,++}|\psi_{E,--}\rangle|^2})$. Now:

$$|\langle\psi_{E,++}|\psi_{E,--}\rangle| = \left(2\sqrt{P_g(1-P_g)}\right)^2 = \eta^2$$

whence $P'_{g,\text{even}} = \frac{1}{2}(1+\sqrt{1-\eta^4})$.

- With probability $\frac{1}{2}$ she announces $c_1 c_2 = -1$. Then Eve knows that she has one of the two states

$$\psi_{E,+-} = \left(\sqrt{P_g}\,|e_1\rangle + \sqrt{1-P_g}\,\Big|e_1^\perp\Big\rangle\right) \otimes \left(\sqrt{1-P_g}\,|e_2\rangle + \sqrt{P_g}\,\Big|e_2^\perp\Big\rangle\right)$$
$$\psi_{E,-+} = \left(\sqrt{1-P_g}\,|e_1\rangle + \sqrt{P_g}\,\Big|e_1^\perp\Big\rangle\right) \otimes \left(\sqrt{P_g}\,|e_2\rangle + \sqrt{1-P_g}\,\Big|e_2^\perp\Big\rangle\right)$$

with equal *a priori* probability. Once again, $|\langle\psi_{E,+-}\rangle|\langle\psi_{E,+-}\rangle| = \eta^2$ and therefore again $P'_{g,\text{odd}} = \frac{1}{2}(1+\sqrt{1-\eta^4})$.

In summary, Eve's probability of guessing the first bit correctly has now increased to

$$P'_g = \frac{1}{2}(1+\sqrt{1-\eta^4}). \tag{G.23}$$

The underlying physical reason is of course that the measurement that achieves the optimal guessing probability is an entangled measurement on both Eve's systems E_1 and E_2. If she were restricted to perform local measurements, her strategy would reduce to one with classical side-information.[3]

G.6 Simulations of the Singlet Behavior

In the main text, we have described how the singlet behavior (9.4)

$$\mathcal{P}_{\Psi^-} = \left\{ P(a,b|\hat{a},\hat{b}) = \frac{1}{4}(1-ab\hat{a}\cdot\hat{b})\,\Big|\,a,b\in\{-1,+1\},\hat{a},\hat{b}\in\mathbb{S}^2 \right\}. \tag{G.24}$$

---

[3] Indeed, by no-signaling, it does not matter *when* Eve performs her measurement: When at the start of the previous subsection we assumed that the initial state was $|\chi_+\rangle|\chi_+\rangle$, this could have been steered by Eve from the state (G.22) by measuring her systems in the computational basis.

can be simulated if in each round the players, alongside with classical information, share a PR-box (section 9.2) or can send one bit of information (section 11.2). In this appendix, we present the constructive protocols in the version of (Degorre *et al.*, 2005), which unifies both approaches and even connects them with Werner's local variable model for mixed states (subsection 3.3.1). The idea is to use the nonlocal resource in such a way that the local variable shared between Alice and Bob, a vector $\vec{\lambda}$ on the unit sphere, ends up distributed according to

$$Q(\vec{\lambda}) = \frac{1}{2\pi} |\hat{a} \cdot \vec{\lambda}|. \tag{G.25}$$

Then, if Alice outputs $a = \operatorname{sign}(\hat{a} \cdot \vec{\lambda})$ and Bob outputs $b = -\operatorname{sign}(\hat{b} \cdot \vec{\lambda})$, one has $\langle a \rangle = \langle b \rangle = 0$ and

$$\langle ab \rangle = -\int_{\mathbb{S}^2} d\vec{\lambda} \frac{1}{2\pi} \underbrace{|\hat{a} \cdot \vec{\lambda}| \operatorname{sign}(\hat{a} \cdot \vec{\lambda})}_{=\hat{a} \cdot \vec{\lambda}} \operatorname{sign}(\hat{b} \cdot \vec{\lambda}) = -\vec{a} \cdot \vec{b}$$

as readily proved by passing in spherical coordinates chosen such that $\hat{a} \equiv \hat{z}$ and $\hat{b} \equiv \cos\beta\hat{z} + \sin\beta\hat{x}$. Clearly, the distribution (G.25) could not have been arranged prior to the round of the test, as it depends on Alice's actual input $\hat{a}$.

G.6.1   Alice's sampling

As a first step, let us see how Alice can generate locally the distribution (G.25). Consider first the "rejection method":

1. Pick $\vec{\lambda}$ uniformly on $\mathbb{S}^2$ and u uniformly in $[0, 1]$;
2. Keep $\vec{\lambda}$ if $|\hat{a} \cdot \vec{\lambda}| \geq u$, discard it otherwise.

The probability that a given $\vec{\lambda}$ is kept is the probability that one draws a value u smaller than $|\hat{a} \cdot \vec{\lambda}|$; since u is drawn uniformly, this probability is just $|\hat{a} \cdot \vec{\lambda}|$. Therefore $Q(\vec{\lambda}) \propto |\hat{a} \cdot \vec{\lambda}|$. By normalizing *a posteriori*, one finds the factor $\frac{1}{2\pi}$. Thus, a $\vec{\lambda}$ chosen with the rejection method is distributed according to (G.25).

The problem with the rejection method, as the name indicates, is that several instances of $\vec{\lambda}$ are discarded (half of them on average). This is solved by noticing that, if $\vec{\lambda}$ is drawn uniformly in $\mathbb{S}^2$, $u \equiv |\hat{a} \cdot \vec{\lambda}|$ is uniform in $[0, 1]$. Assume then that Alice starts with *two* unit vectors $\vec{\lambda}_0$ and $\vec{\lambda}_1$, drawn independently, each with uniform distribution on the unit sphere $\mathbb{S}^2$. She can then sample her $\vec{\lambda}$ according to this rule (*"choice method"*): If $|\hat{a} \cdot \vec{\lambda}_0| \geq |\hat{a} \cdot \vec{\lambda}_1|$, set $\vec{\lambda} = \vec{\lambda}_0$; otherwise, set $\vec{\lambda} = \vec{\lambda}_1$. Now, whenever $\vec{\lambda} = \vec{\lambda}_0$ is selected, it is using the rejection method with $u_0 \equiv |\hat{a} \cdot \vec{\lambda}_1|$; whenever $\vec{\lambda} = \vec{\lambda}_1$ is selected, it is using the rejection method with $u_1 \equiv |\hat{a} \cdot \vec{\lambda}_0|$. Therefore, in each round the $\vec{\lambda}$ chosen by Alice is distributed according to (G.25).

G.6.2   Application to simulation

Let us now suppose that, prior to each round, Alice and Bob share two unit vectors $\vec{\lambda}_0$ and $\vec{\lambda}_1$, drawn independently and uniformly on the sphere. The simulation of the singlet behavior with *one bit of communication* is trivial: Alice picks uses the choice method to pick one of the two vectors and tells Bob which one she has chosen; Bob uses the same one to produce his output.[4]

The simulation with *one use of the PR-box* is only slightly more involved. First notice that which λ_j Bob uses does not matter if $\text{sign}(\hat{b} \cdot \vec{\lambda}_0) = \text{sign}(\hat{b} \cdot \vec{\lambda}_1)$. So, if by default Bob outputs $b = -\text{sign}(\hat{b} \cdot \vec{\lambda}_0)$, he has to flip his output only when *both* $\vec{\lambda}_1$ is chosen by Alice and $\text{sign}(\hat{b} \cdot \vec{\lambda}_0) = -\text{sign}(\hat{b} \cdot \vec{\lambda}_1)$. Consider then the following protocol:

- Alice runs the choice method and chooses $\vec{\lambda} = \vec{\lambda}_j$. She inputs $x = j$ in the PR-box and receives the output $\alpha \in \{0, 1\}$. Finally, Alice outputs $a = \text{sign}((-1)^\alpha \hat{a} \cdot \vec{\lambda})$.
- Bob checks if $\text{sign}(\hat{b} \cdot \vec{\lambda}_0) = \text{sign}(\hat{b} \cdot \vec{\lambda}_1)$: If it is, he inputs $y = 0$ in the PR-box; otherwise, he inputs $y = 1$ and receives the output $\beta \in \{0, 1\}$. Finally, Bob outputs $b = -\text{sign}((-1)^\beta \hat{b} \cdot \vec{\lambda}_0)$.

The conditions for $xy = 1$ are exactly the conditions under which Bob's bit should be flipped in the previous narrative. Here, when $xy = 1$ either player flips the output but not both; whereas in the rounds where $xy = 0$, either both players keep their output or both flip it. This is a necessity of the PR-box being no-signaling: Bob should not get to know if $\vec{\lambda}_1$ was chosen.

Interestingly, the case where Alice samples according to the choice method but *no additional nonlocal resources are available* is also instructive, because it connects with other considerations. One can imagine that Alice outputs $a = \text{sign}(\hat{a} \cdot \vec{\lambda})$ with the $\vec{\lambda}$ chosen with the choice method, while Bob outputs $b = -\text{sign}(\hat{b} \cdot \vec{\lambda}_0)$. This will produce the correct correlations in half of the rounds, while in the other half the oucomes will be uncorrelated (because Alice has used $\vec{\lambda}_1$ that is independent of $\vec{\lambda}_0$). The resulting behavior, which must be local, is nothing else than that of the Werner state with $W = \frac{1}{2}$. Alternatively, one can imagine that, knowing that Bob is going to use $\vec{\lambda}_0$, Alice refuses to produce an output in the rounds when she should have used $\vec{\lambda}_1$. The post-selected behavior is the singlet behavior. This is an example of simulation exploiting the detection loophole, in which Alice has 50% efficiency (we saw in subsection 1.5.2 that with those efficiencies one could actually reach $S = 4$).

G.7 Properties of the Variational Distance

The proofs of subsections 9.4.2 and 11.5.3 use the notion of variational distance between two probability distributions on the same alphabet

$$\mathcal{D}[P_C, Q_C] = \frac{1}{2} \sum_{c \in C} |P_C(c) - Q_C(c)|. \tag{G.26}$$

Here we prove three properties that are used in those derivations:

---

[4] Notice that, in this protocol, Alice's communication cannot be compressed over several runs, because the bit is unbiased. This shows that this protocol is not identical to the one of Toner and Bacon (2003), where some bias in the communication could be exploited to compress the average information down to approximately 0.85 bits per round.

- The triangle inequality follows immediately from $|a+b| \le |a| + |b|$: Indeed,

$$\mathcal{D}[P,R] = \frac{1}{2}\sum_c |P(c) - R(c)| = \frac{1}{2}\sum_c |(P(c) - Q(c)) + (Q(c) - R(c))|$$

$$\le \frac{1}{2}\sum_c (|P(c) - Q(c)| + |Q(c) - R(c)|) = \mathcal{D}[P,Q] + \mathcal{D}[Q,R].$$

Thus, $\mathcal{D}$ is indeed a distance, the three other requirements being obviously satisfied: $\mathcal{D}[P,Q] \ge 0$, $\mathcal{D}[P,Q] = 0$ if and only if $P = Q$ and $\mathcal{D}[P,Q] = \mathcal{D}[Q,P]$.

- Given a joint probability distribution P_{AB} on two identical alphabets $\mathcal{A} = \mathcal{B}$, the distance between the marginal distributions is bounded by $\mathcal{D}[P_A, P_B] \le P_{AB}(a \ne b)$. Indeed,

$$\mathcal{D}[P_A, P_B] = \frac{1}{2}\sum_{\xi} |P(a = \xi) - P(b = \xi)|$$

$$\overset{*}{=} \frac{1}{2}\sum_{\xi} \left| \sum_{b \ne \xi} P(\xi, b) - \sum_{a \ne \xi} P(a, \xi) \right|$$

$$\le \frac{1}{2}\left(\sum_{\xi}\sum_{b \ne \xi} P(\xi, b) + \sum_{\xi}\sum_{a \ne \xi} P(a, \xi) \right)$$

$$= P(a \ne b)$$

where in the starred equality we have used $P(a = \xi) = P(\xi, \xi) + \sum_{b \ne \xi} P(\xi, b)$ and $P(b = \xi) = P(\xi, \xi) + \sum_{a \ne \xi} P(a, \xi)$.

- For bits $\mathcal{A} = \{0, 1\}$, it holds $\mathcal{D}[\tilde{P}_A, P_A] = 2\mathcal{D}[P_A, Q_A]$, where $\tilde{P}_A$ is defined by $\tilde{P}(a) = 1 - P(a) = P(1 - a)$ and $Q_A(a) = \frac{1}{2}$. Indeed,

$$\mathcal{D}[\tilde{P}_A, P_A] = \frac{1}{2}\sum_{\xi} |P(1 - a = \xi) - P(a = \xi)|$$

$$= \frac{1}{2}\sum_{\xi} |1 - 2P(a = \xi)|$$

$$= 2\left[\frac{1}{2}\sum_{\xi} \left| \frac{1}{2} - P(a = \xi) \right| \right] = 2\mathcal{D}[P_A, Q_A].$$

G.8   Information Causality from Desiderata on Information Entropies

We have mentioned in subsection 10.2.3 that the condition (10.3) for Information Causality (IC) holds under basic requirements on the information entropy of the underlying theory. For the convenience of the reader, we reproduce here the proof of (Al-Safi and Short, 2011) with our notation and some additional comments.

The two requirements on the information entropy H of the theory are:

1. Consistency with classical information theory: If A is a resource that can be described in classical theory, then

$$H(\mathbf{A}) = H_c(\mathbf{A}) \tag{G.27}$$

   where H_c the Shannon entropy for classical variables.

2. A form of the data-processing inequality: If $B \to B'$ denotes a transformation on Bob's system alone, it must hold

$$H(\mathbf{A}|\mathbf{B}') \geq H(\mathbf{A}|\mathbf{B}) \tag{G.28}$$

   where $H(\mathbf{A}|\mathbf{B}) = H(\mathbf{A}, \mathbf{B}) - H(\mathbf{B})$. Using this expression, it is useful to consider the case where B' consists in simply discarding the system: In this case, (G.28) becomes $H(\mathbf{A}) \geq H(\mathbf{A}, \mathbf{B}) - H(\mathbf{B})$, when subadditivity follows:

$$H(\mathbf{A}, \mathbf{B}) \leq H(\mathbf{A}) + H(\mathbf{B}). \tag{G.29}$$

Now we can present the proof. The resource of Bob is denoted by ρ_B with a hint to the quantum formalizm: But this is purely a notation, quantum theory is not assumed here.

Since Bob's guess β_y is obtained from processing his resource ρ_B and Alice's message $\vec{m} \in \mathbf{M} = \{0, 1\}^\kappa$, it holds

$$H_c(\mathbf{X}_y|\beta_y) \overset{(G.28)}{\geq} H(\mathbf{X}_y|\rho_B, \mathbf{M}). \tag{G.30}$$

Notice that there is no distribution on ρ_B, since this is the resource the players are using; but it might have an entropy, and this is why on the r.h.s. we have to work with the information entropy of the theory. Now we denote $\vec{\mathbf{X}} = (\mathbf{X}_0, \dots, \mathbf{X}_{N-1})$ the collection of all Alice's inputs. The following bounds hold

$$\sum_y H(\mathbf{X}_y|\rho_B, \mathbf{M}) \overset{(G.29)}{\geq} H(\vec{\mathbf{X}}|\rho_B, \mathbf{M}) = H(\vec{\mathbf{X}}, \rho_B, \mathbf{M}) - H(\rho_B, \mathbf{M})$$

$$\overset{(G.29)}{\geq} H(\vec{\mathbf{X}}, \rho_B, \mathbf{M}) - H(\rho_B) - H_c(\mathbf{M})$$

$$\overset{*}{=} H(\vec{\mathbf{X}}, \rho_B, \mathbf{M}) - H(\vec{\mathbf{X}}, \rho_B) + H_c(\vec{\mathbf{X}}) - H_c(\mathbf{M})$$

$$= H(\mathbf{M}|\vec{\mathbf{X}}, \rho_B) + H_c(\vec{\mathbf{X}}) - H_c(\mathbf{M})$$

$$\geq H_c(\vec{\mathbf{X}}) - \kappa$$

where for the step * we used the fact that Bob's resource must be uncorrelated with Alice's inputs; for the last step, we used the facts that the conditional entropy is positive and that $H_c(\mathbf{M}) \leq \kappa$ because Alice sends those many bits.

This bound is already a form of IC that does not assume the distribution of Alice's inputs. If we add that her inputs must be uncorrelated, as we assumed in the main text, then $H_c(\vec{\mathbf{X}}) = \sum_y H_c(\mathbf{X}_y)$ and we recover (10.3).

References

Acín, A., Andrianov, A., Jané, E., and Tarrach, R. (2001). Three-qubit pure-state canonical forms. *Journal of Physics A: Mathematical and General*, **34** (35), 6725.

Acín, Antonio, Brunner, Nicolas, Gisin, Nicolas, Massar, Serge, Pironio, Stefano, and Scarani, Valerio (2007, Jun). Device-independent security of quantum cryptography against collective attacks. *Phys. Rev. Lett.*, **98**, 230501.

Acín, A., Durt, T., Gisin, N., and Latorre, J. I. (2002, May). Quantum nonlocality in two three-level systems. *Phys. Rev. A*, **65**, 052325.

Acín, Antonio, Fritz, Tobias, Leverrier, Anthony, and Sainz, Ana Belén (2015, Mar). A combinatorial approach to nonlocality and contextuality. *Communications in Mathematical Physics*, **334** (2), 533–628.

Acín, Antonio, Gisin, Nicolas, and Masanes, Lluis (2006, Sep). From Bell's theorem to secure quantum key distribution. *Phys. Rev. Lett.*, **97**, 120405.

Acín, Antonio and Masanes, Lluis (2016, Dec). Certified randomness in quantum physics. *Nature*, **540**, 213.

Acín, Antonio, Massar, Serge, and Pironio, Stefano (2012, Mar). Randomness versus nonlocality and entanglement. *Phys. Rev. Lett.*, **108**, 100402.

Acín, Antonio, Pironio, Stefano, Vértesi, Tamás, and Wittek, Peter (2016, Apr). Optimal randomness certification from one entangled bit. *Phys. Rev. A*, **93**, 040102.

Al-Safi, Sabri W. and Short, Anthony J. (2011, Oct). Information causality from an entropic and a probabilistic perspective. *Phys. Rev. A*, **84**, 042323.

Allcock, Jonathan, Brunner, Nicolas, Pawlowski, Marcin, and Scarani, Valerio (2009, Oct). Recovering part of the boundary between quantum and nonquantum correlations from information causality. *Phys. Rev. A*, **80**, 040103.

Almeida, Mafalda L., Bancal, Jean-Daniel, Brunner, Nicolas, Acín, Antonio, Gisin, Nicolas, and Pironio, Stefano (2010, Jun). Guess your neighbor's input: A multipartite nonlocal game with no quantum advantage. *Phys. Rev. Lett.*, **104**, 230404.

Aravind, P. K. (2002, Aug). Bell's theorem without inequalities and only two distant observers. *Found. Phys. Lett.*, **15** (4), 397–405.

Ardehali, M. (1992, Nov). Bell inequalities with a magnitude of violation that grows exponentially with the number of particles. *Phys. Rev. A*, **46**, 5375–5378.

Arnon-Friedman, Rotem, Dupuis, Frédéric, Fawzi, Omar, Renner, Renato, and Vidick, Thomas (2018). Practical device-independent quantum cryptography via entropy accumulation. *Nature Communications*, **9** (1), 459.

Arnon-Friedman, Rotem and Renner, Renato (2015). De Finetti reductions for correlations. *Journal of Mathematical Physics*, **56** (5), 052203.

Arnon-Friedman, Rotem, Renner, Renato, and Vidick, Thomas (2019). Simple and tight device-independent security proofs. SIAM Journal on Computing 48 (1), 181.

Arnon-Friedman, Rotem and Ta-Shma, Amnon (2012, Dec). Limits of privacy amplification against nonsignaling memory attacks. *Phys. Rev. A*, **86**, 062333.

Arora, Sanjeev and Barak, Joaz (2009). *Computational Complexity: A Modern Approach*. Cambridge University Press, Cambridge.

Aspect, Alain, Dalibard, Jean, and Roger, Gérard (1982*a*, Dec). Experimental test of Bell's inequalities using time-varying analyzers. *Phys. Rev. Lett.*, **49**, 1804–1807.

Aspect, Alain, Grangier, Philippe, and Roger, Gérard (1982*b*, Jul). Experimental realization of Einstein-Podolsky-Rosen-Bohm gedankenexperiment: A new violation of Bell's inequalities. *Phys. Rev. Lett.*, **49**, 91–94.

Augusiak, R., Demianowicz, M., and Acín, A. (2014). Local hidden-variable models for entangled quantum states. *Journal of Physics A: Mathematical and Theoretical*, **47** (42), 424002.

Augusiak, Remigiusz and Horodecki, Pawel (2006, Jul). Bound entanglement maximally violating Bell inequalities: Quantum entanglement is not fully equivalent to cryptographic security. *Phys. Rev. A*, **74**, 010305.

Bamps, Cédric and Pironio, Stefano (2015, May). Sum-of-squares decompositions for a family of clauser-horne-shimony-holt-like inequalities and their application to self-testing. *Phys. Rev. A*, **91**, 052111.

Banaszek, Konrad and Wódkiewicz, Krzysztof (1998, Dec). Nonlocality of the Einstein-Podolsky-Rosen state in the Wigner representation. *Phys. Rev. A*, **58**, 4345–4347.

Bancal, Jean-Daniel, Barrett, Jonathan, Gisin, Nicolas, and Pironio, Stefano (2013, Jul). Definitions of multipartite nonlocality. *Phys. Rev. A*, **88**, 014102.

Bancal, Jean-Daniel, Brunner, Nicolas, Gisin, Nicolas, and Liang, Yeong-Cherng (2011*a*, Jan). Detecting genuine multipartite quantum nonlocality: A simple approach and generalization to arbitrary dimensions. *Phys. Rev. Lett.*, **106**, 020405.

Bancal, Jean-Daniel, Gisin, Nicolas, Liang, Yeong-Cherng, and Pironio, Stefano (2011*b*, Jun). Device-independent witnesses of genuine multipartite entanglement. *Phys. Rev. Lett.*, **106**, 250404.

Bancal, Jean-Daniel, Navascués, Miguel, Scarani, Valerio, Vértesi, Tamás, and Yang, Tzyh Haur (2015, Feb). Physical characterization of quantum devices from nonlocal correlations. *Phys. Rev. A*, **91**, 022115.

Bancal, J-D., Pironio, S., Acin, A., Liang, Y-C., Scarani, V., and Gisin, N. (2012, Dec). Quantum non-locality based on finite-speed causal influences leads to superluminal signalling. *Nat. Phys.*, **8** (12), 867–870.

Bancal, Jean-Daniel, Sheridan, Lana, and Scarani, Valerio (2014). More randomness from the same data. *New Journal of Physics*, **16** (3), 033011.

Barnea, Tomer Jack, Bancal, Jean-Daniel, Liang, Yeong-Cherng, and Gisin, Nicolas (2013, Aug). Tripartite quantum state violating the hidden-influence constraints. *Phys. Rev. A*, **88**, 022123.

Barrett, Jonathan (2002, Mar). Nonsequential positive-operator-valued measurements on entangled mixed states do not always violate a Bell inequality. *Phys. Rev. A*, **65**, 042302.

Barrett, Jonathan (2007, Mar). Information processing in generalized probabilistic theories. *Phys. Rev. A*, **75**, 032304.

Barrett, Jonathan, Colbeck, Roger, and Kent, Adrian (2013, Jan). Memory attacks on device-independent quantum cryptography. *Phys. Rev. Lett.*, **110**, 010503.

Barrett, Jonathan, Collins, Daniel, Hardy, Lucien, Kent, Adrian, and Popescu, Sandu (2002, Oct). Quantum nonlocality, Bell inequalities, and the memory loophole. *Phys. Rev. A*, **66**, 042111.

Barrett, Jonathan and Gisin, Nicolas (2011, Mar). How much measurement independence is needed to demonstrate nonlocality? *Phys. Rev. Lett.*, **106**, 100406.

Barrett, Jonathan, Hardy, Lucien, and Kent, Adrian (2005*a*, Jun). No signaling and quantum key distribution. *Phys. Rev. Lett.*, **95**, 010503.

Barrett, Jonathan, Kent, Adrian, and Pironio, Stefano (2006, Oct). Maximally nonlocal and monogamous quantum correlations. *Phys. Rev. Lett.*, **97**, 170409.

Barrett, Jonathan, Linden, Noah, Massar, Serge, Pironio, Stefano, Popescu, Sandu, and Roberts, David (2005*b*, Feb). Nonlocal correlations as an information-theoretic resource. *Phys. Rev. A*, **71**, 022101.

Barrett, Jonathan and Pironio, Stefano (2005, Sep). Popescu-rohrlich correlations as a unit of nonlocality. *Phys. Rev. Lett.*, **95**, 140401.

Belinsky, A. V. and Klyshko, D. N. (1993*a*). A modified n-particle Bell theorem, the corresponding optical experiment and its classical model. *Physics Letters A*, **176** (6), 415–420.

Belinsky, A. V. and Klyshko, D. N. (1993*b*). Interference of light and Bell's theorem. *Physics-Uspekhi*, **36** (8), 653.

Bell, John Stuart (1964). On the Einstein-Podolski-Rosen paradox. *Physics*, **1**, 195.

Bell, John S. (1966, Jul). On the problem of hidden variables in quantum mechanics. *Rev. Mod. Phys.*, **38**, 447–452.

Bell, John S. (2004). *Speakable and unspeakable in quantum mechanics*. Cambridge University Press, Cambridge.

Bendersky, Ariel, de la Torre, Gonzalo, Senno, Gabriel, Figueira, Santiago, and Acín, Antonio (2016, Jun). Algorithmic pseudorandomness in quantum setups. *Phys. Rev. Lett.*, **116**, 230402.

Bene, Erika and Vértesi, Tamás (2018). Measurement incompatibility does not give rise to Bell violation in general. *New Journal of Physics*, **20** (1), 013021.

Bennett, Charles H. and Brassard, Gilles (1984). Quantum cryptography: Public key distribution and coin tossing. In *Proceedings of IEEE International Conference on Computers, Systems, and Signal Processing*, pp. 175–179. IEEE, New York.

Bennett, Charles H., Brassard, Gilles, and Mermin, N. David (1992, Feb). Quantum cryptography without Bell's theorem. *Phys. Rev. Lett.*, **68**, 557–559.

Bierhorst, Peter, Knill, Emanuel, Glancy, Scott, Zhang, Yanbao, Mink, Alan, Jordan, Stephen, Rommal, Andrea, Liu, Yi-Kai, Christensen, Bradley, Nam, Sae Woo, Stevens, Martin J., and Shalm, Lynden K. (2018). Experimentally generated randomness certified by the impossibility of superluminal signals. *Nature*, **556** (7700), 223–226.

Blass, Andreas and Gurevich, Yuri (2017). Common denominator for value and expectation no-go theorems. Preprint arXiv:1707.07368.

Bowles, Joseph, Šupić, Ivan, Cavalcanti, Daniel, and Acín, Antonio (2018, Oct). Self-testing of Pauli observables for device-independent entanglement certification. *Phys. Rev. A*, **98**, 042336.

Boyd, Stephen and Vandenberghe, Lieven (2004). *Convex Optimization*. Cambridge University Press, Cambridge.

Branciard, Cyril, Brunner, Nicolas, Buhrman, Harry, Cleve, Richard, Gisin, Nicolas, Portmann, Samuel, Rosset, Denis, and Szegedy, Mario (2012*a*, Sep). Classical simulation of entanglement swapping with bounded communication. *Phys. Rev. Lett.*, **109**, 100401.

Branciard, Cyril, Brunner, Nicolas, Gisin, Nicolas, Kurtsiefer, Christian, Lamas-Linares, Antia, Ling, Alexander, and Scarani, Valerio (2008, Jul). Testing quantum correlations versus single-particle properties within Leggett's model and beyond. *Nature Physics*, **4**, 681.

Branciard, Cyril, Rosset, Denis, Gisin, Nicolas, and Pironio, Stefano (2012b, Mar). Bilocal versus nonbilocal correlations in entanglement-swapping experiments. *Phys. Rev. A*, **85**, 032119.

Branciard, Cyril, Rosset, Denis, Liang, Yeong-Cherng, and Gisin, Nicolas (2013, Feb). Measurement-device-independent entanglement witnesses for all entangled quantum states. *Phys. Rev. Lett.*, **110**, 060405.

Brassard, Gilles, Buhrman, Harry, Linden, Noah, Méthot, André Allan, Tapp, Alain, and Unger, Falk (2006, Jun). Limit on nonlocality in any world in which communication complexity is not trivial. *Phys. Rev. Lett.,* **96**, 250401.

Brassard, Gilles, Devroye, Luc, and Gravel, Claude (2019). Remote sampling with applications to general entanglement simulation. Entropy **21**, 92.

Braunstein, Samuel L. and Caves, Carlton M. (1990). Wringing out better Bell inequalities. *Annals of Physics*, **202** (1), 22–56.

Braunstein, Samuel L., Mann, A., and Revzen, M. (1992, Jun). Maximal violation of Bell inequalities for mixed states. *Phys. Rev. Lett.,* **68**, 3259–3261.

Brunner, Nicolas, Cavalcanti, Daniel, Pironio, Stefano, Scarani, Valerio, and Wehner, Stephanie (2014, Apr). Bell nonlocality. *Rev. Mod. Phys.,* **86**, 419–478.

Brunner, Nicolas, Gisin, Nicolas, and Scarani, Valerio (2005). Entanglement and non-locality are different resources. *New Journal of Physics*, **7** (1), 88.

Brunner, Nicolas, Pironio, Stefano, Acín, Antonio, Gisin, Nicolas, Méthot, André Allan, and Scarani, Valerio (2008, May). Testing the dimension of hilbert spaces. *Phys. Rev. Lett.,* **100**, 210503.

Bub, Jeffrey (2015). *Bananaworld: Quantum Mechanics for Primates*. Oxford University Press, Oxford.

Buhrman, Harry, Cleve, Richard, Massar, Serge, and de Wolf, Ronald (2010, Mar). Nonlocality and communication complexity. *Rev. Mod. Phys.,* **82**, 665–698.

Buscemi, Francesco (2012, May). All entangled quantum states are nonlocal. *Phys. Rev. Lett.,* **108**, 200401.

Cabello, Adán (2001a, Jun). "All versus nothing" inseparability for two observers. *Phys. Rev. Lett.,* **87**, 010403.

Cabello, Adán (2001b, Mar). Bell's theorem without inequalities and without probabilities for two observers. *Phys. Rev. Lett.,* **86**, 1911–1914.

Cabello, Adán (2008, Nov). Experimentally testable state-independent quantum contextuality. *Phys. Rev. Lett.,* **101**, 210401.

Cabello, Adán (2013, Feb). Simple explanation of the quantum violation of a fundamental inequality. *Phys. Rev. Lett.,* **110**, 060402.

Cabello, Adán, Kleinmann, Matthias, and Portillo, José R (2016). Quantum state-independent contextuality requires 13 rays. *Journal of Physics A: Mathematical and Theoretical*, **49** (38), 38LT01.

Cabello, Adán, Severini, Simone, and Winter, Andreas (2010). (Non-)Contextuality of physical theories as an axiom. Preprint arXiv:1010.2163.

Cabello, Adán, Severini, Simone, and Winter, Andreas (2014, Jan). Graph-theoretic approach to quantum correlations. *Phys. Rev. Lett.,* **112**, 040401.

Cavalcanti, Daniel, Salles, Alejo, and Scarani, Valerio (2010, Dec). Macroscopically local correlations can violate information causality. *Nature Communications*, **1**, 136 EP –.

Cavalcanti, Daniel and Skrzypczyk, Paul (2017). Quantum steering: A review with focus on semidefinite programming. *Reports on Progress in Physics*, **80** (2), 024001.

Cavalcanti, Eric G., Hall, Michael J. W., and Wiseman, Howard M. (2013, Mar). Entanglement verification and steering when alice and bob cannot be trusted. *Phys. Rev. A*, **87**, 032306.

Cerf, N. J., Gisin, N., Massar, S., and Popescu, S. (2005, Jun). Simulating maximal quantum entanglement without communication. *Phys. Rev. Lett.,* **94**, 220403.

Chaves, Rafael, Cavalcanti, Daniel, and Aolita, Leandro (2017). Causal hierarchy of multipartite Bell nonlocality. *Quantum*, **1**, 23.

Chaves, R., Kueng, R., Brask, J. B., and Gross, D. (2015, Apr). Unifying framework for relaxations of the causal assumptions in Bell's theorem. *Phys. Rev. Lett.*, **114**, 140403.

Cirel'son, B. S. (1980, Mar). Quantum generalizations of Bell's inequality. *Letters in Mathematical Physics*, **4** (2), 93–100.

Clauser, John F. and Horne, Michael A. (1974, Jul). Experimental consequences of objective local theories. *Phys. Rev. D*, **10**, 526–535.

Clauser, John F., Horne, Michael A., Shimony, Abner, and Holt, Richard A. (1969, Oct). Proposed experiment to test local hidden-variable theories. *Phys. Rev. Lett.*, **23**, 880–884.

Cleve, R., Hoyer, P., Toner, B., and Watrous, J. (2004, Jun). Consequences and limits of nonlocal strategies. In *Proceedings. 19th IEEE Annual Conference on Computational Complexity, 2004.*, pp. 236–249.

Coladangelo, Andrea, Goh, Koon Tong, and Scarani, Valerio (2017a, May). All pure bipartite entangled states can be self-tested. *Nature Communications*, **8**, 15485.

Coladangelo, Andrea, Grilo, Alex, Jeffery, Stacey, and Vidick, Thomas (2017b). Verifier-on-a-leash: New schemes for verifiable delegated quantum computation, with quasilinear resources. Preprint arXiv:1708.07359.

Coladangelo, Andrea and Stark, Jalex (2018). Unconditional separation of finite and infinite-dimensional quantum correlations. Preprint arXiv:1804.05116.

Colbeck, Roger and Kent, Adrian (2011). Private randomness expansion with untrusted devices. *Journal of Physics A: Mathematical and Theoretical*, **44** (9), 095305.

Colbeck, Roger and Renner, Renato (2008, Aug). Hidden variable models for quantum theory cannot have any local part. *Phys. Rev. Lett.*, **101**, 050403.

Colbeck, Roger and Renner, Renato (2012, May). Free randomness can be amplified. *Nature Physics*, **8**, 450.

Collins, Daniel and Gisin, Nicolas (2004). A relevant two qubit Bell inequality inequivalent to the chsh inequality. *J. Phys. A: Math. Theor.*, **37**, 1775.

Collins, Daniel, Gisin, Nicolas, Linden, Noah, Massar, Serge, and Popescu, Sandu (2002a, Jan). Bell inequalities for arbitrarily high-dimensional systems. *Phys. Rev. Lett.*, **88**, 040404.

Collins, Daniel, Gisin, Nicolas, Popescu, Sandu, Roberts, David, and Scarani, Valerio (2002b, Apr). Bell-type inequalities to detect true n-body nonseparability. *Phys. Rev. Lett.*, **88**, 170405.

Cong, Wan, Cai, Yu, Bancal, Jean-Daniel, and Scarani, Valerio (2017, Aug). Witnessing irreducible dimension. *Phys. Rev. Lett.*, **119**, 080401.

Coretti, Sandro, Hänggi, Esther, and Wolf, Stefan (2011, Aug). Nonlocality is transitive. *Phys. Rev. Lett.*, **107**, 100402.

Curchod, F. J., Johansson, M., Augusiak, R., Hoban, M. J., Wittek, P., and Acín, A. (2017, Feb). Unbounded randomness certification using sequences of measurements. *Phys. Rev. A*, **95**, 020102.

Dakić, Borivoje, Šuvakov, Milovan, Paterek, Tomasz, and Brukner, Časlav (2008, Nov). Efficient hidden-variable simulation of measurements in quantum experiments. *Phys. Rev. Lett.*, **101**, 190402.

Dall'Arno, Michele, Brandsen, Sarah, Tosini, Alessandro, Buscemi, Francesco, and Vedral, Vlatko (2017, Jul). No-hypersignaling principle. *Phys. Rev. Lett.*, **119**, 020401.

D'Ariano, Giacomo Mauro, Chiribella, Giulio, and Perinotti, Paolo (2017). *Quantum Theory from First Principles*. Cambridge University Press, Cambridge.

De, A., Portmann, C., Vidick, T., and Renner, R. (2012). Trevisan's extractor in the presence of quantum side information. *SIAM Journal on Computing*, **41** (4), 915–940.

de la Torre, Gonzalo, Hoban, Matty J., Dhara, Chirag, Prettico, Giuseppe, and Acín, Antonio (2015, Apr). Maximally nonlocal theories cannot be maximally random. *Phys. Rev. Lett.*, **114**, 160502.

Degorre, Julien, Kaplan, Marc, Laplante, Sophie, and Roland, Jérémie (2009). The communication complexity of non-signaling distributions. In *Mathematical Foundations of Computer Science 2009* (ed. R. Královič and D. Niwiński), pp. 270–281. Springer, Berlin, Heidelberg.

Degorre, Julien, Laplante, Sophie, and Roland, Jérémie (2005, Dec). Simulating quantum correlations as a distributed sampling problem. *Phys. Rev. A*, **72**, 062314.

Donohue, John Matthew and Wolfe, Elie (2015, Dec). Identifying nonconvexity in the sets of limited-dimension quantum correlations. *Phys. Rev. A*, **92**, 062120.

Duarte, Cristhiano, Brito, Samuraí, Amaral, Barbara, and Chaves, Rafael (2018, Dec). Concentration phenomena in the geometry of Bell correlations. *Phys. Rev. A*, **98**, 062114.

Eberhard, Philippe H. (1993, Feb). Background level and counter efficiencies required for a loophole-free Einstein-Podolsky-Rosen experiment. *Phys. Rev. A*, **47**, R747–R750.

Einstein, A., Podolsky, B., and Rosen, N. (1935, May). Can quantum-mechanical description of physical reality be considered complete? *Phys. Rev.*, **47**, 777–780.

Ekert, Artur K. (1991, Aug). Quantum cryptography based on Bell's theorem. *Phys. Rev. Lett.*, **67**, 661–663.

Elitzur, Avshalom C., Popescu, Sandu, and Rohrlich, Daniel (1992). Quantum nonlocality for each pair in an ensemble. *Physics Letters A*, **162** (1), 25–28.

Elkouss, David and Wehner, Stephanie (2016). (Nearly) optimal P values for all Bell inequalities. *npj Quantum Inf* **2**, 16026.

Fine, Arthur (1982, Feb). Hidden variables, joint probability, and the bell inequalities. *Phys. Rev. Lett.*, **48**, 291–295.

Franson, J. D. (1989, May). Bell inequality for position and time. *Phys. Rev. Lett.*, **62**, 2205–2208.

Fraser, Thomas C. and Wolfe, Elie (2018, Aug). Causal compatibility inequalities admitting quantum violations in the triangle structure. *Phys. Rev. A*, **98**, 022113.

Fritz, Tobias (2012). Polyhedral duality in Bell scenarios with two binary observables. *Journal of Mathematical Physics*, **53** (7), 072202.

Fritz, T., Sainz, A. B., Augusiak, R., Brask, Bohr J., Chaves, R., Leverrier, A., and Acín, A. (2013, Aug). Local orthogonality as a multipartite principle for quantum correlations. *Nature Communications*, **4**, 2263.

Froissart, M. (1981, Aug). Constructive generalization of Bell's inequalities. *Il Nuovo Cimento B (1971-1996)*, **64** (2), 241–251.

Fuchs, Christopher A., Mermin, N. David, and Schack, Rüdiger (2014). An introduction to qbism with an application to the locality of quantum mechanics. *American Journal of Physics*, **82** (8), 749–754.

Gachechiladze, Mariami and Guehne, Otfried (2017). Completing the proof of generic quantum nonlocality. *Physics Letters A*, **381** (15), 1281—1285.

Gallego, Rodrigo, Masanes, Lluis, De La Torre, Gonzalo, Dhara, Chirag, Aolita, Leandro, and Acín, Antonio (2013, Oct). Full randomness from arbitrarily deterministic events. *Nature Communications*, **4**, 2654.

Gallego, Rodrigo, Würflinger, Lars Erik, Acín, Antonio, and Navascués, Miguel (2011, Nov). Quantum correlations require multipartite information principles. *Phys. Rev. Lett.*, **107**, 210403.

Gallego, Rodrigo, Würflinger, Lars Erik, Acín, Antonio, and Navascués, Miguel (2012, Aug). Operational framework for nonlocality. *Phys. Rev. Lett.*, **109**, 070401.

García-Patrón, R., Fiurášek, J., Cerf, N. J., Wenger, J., Tualle-Brouri, R., and Grangier, Ph. (2004, Sep). Proposal for a loophole-free Bell test using homodyne detection. *Phys. Rev. Lett.*, **93**, 130409.

Genovese, Marco (2005). Research on hidden variable theories: A review of recent progresses. *Physics Reports*, **413**, 319–396.

Gerhardt, Ilja, Liu, Qin, Lamas-Linares, Antía, Skaar, Johannes, Scarani, Valerio, Makarov, Vadim, and Kurtsiefer, Christian (2011, Oct). Experimentally faking the violation of Bell's inequalities. *Phys. Rev. Lett.*, **107**, 170404.

Gill, Richard D. (2014, Nov). Statistics, causality and Bell's theorem. *Statist. Sci.*, **29** (4), 512–528.

Gisin, Nicolas (1991). Bell's inequality holds for all non-product states. *Physics Letters A*, **154**, 201–202.

Gisin, Nicolas (2011, Feb). Impossibility of covariant deterministic nonlocal hidden-variable extensions of quantum theory. *Phys. Rev. A*, **83**, 020102.

Gisin, Nicolas (2012, Jan). Non-realism: Deep thought or a soft option? *Foundations of Physics*, **42** (1), 80–85.

Gisin, Nicolas (2014). *Quantum Chance*. Springer, Dordrecht.

Gisin, Nicolas, Méthot, André Allan, and Scarani, Valerio (2007). Pseudo-telepathy: Input cardinality and Bell-type inequalities. *International Journal of Quantum Information*, **05** (04), 525–534.

Giustina, Marissa, Versteegh, Marijn A. M., Wengerowsky, Sören, Handsteiner, Johannes, Hochrainer, Armin, Phelan, Kevin, Steinlechner, Fabian, Kofler, Johannes, Larsson, Jan-oAke, Abellán, Carlos, Amaya, Waldimar, Pruneri, Valerio, Mitchell, Morgan W., Beyer, Jörn, Gerrits, Thomas, Lita, Adriana E., Shalm, Lynden K., Nam, Sae Woo, Scheidl, Thomas, Ursin, Rupert, Wittmann, Bernhard, and Zeilinger, Anton (2015, Dec). Significant-loophole-free test of Bell's theorem with entangled photons. *Phys. Rev. Lett.*, **115**, 250401.

Goh, Koon Tong, Kaniewski, Jedrzej, Wolfe, Elie, Vértesi, Tamás, Wu, Xingyao, Cai, Yu, Liang, Yeong-Cherng, and Scarani, Valerio (2018, Feb). Geometry of the set of quantum correlations. *Phys. Rev. A*, **97**, 022104.

Greenberger, Daniel M., Horne, Michael, and Zeilinger, Anton (1989). Going beyond Bell's theorem. In *Bell's Theorem, Quantum Theory, and Conceptions of the Universe* (ed. M. Kafatos), pp. 69–72. Kluwer Academic, Dordrecht.

Griffith, Meghan (2013). *Free Will: The Basics*. Routledge, Abingdon.

Gu, Yanwu, Li, Weijun, Evans, Michael, and Englert, Berthold-Georg (2019). Very strong evidence in favor of quantum mechanics and against local hidden variables from a bayesian analysis. Preprint Phys. Rev. A **99**, 022112.

Gühne, Otfried, Tóth, Géza, Hyllus, Philipp, and Briegel, Hans J. (2005, Sep). Bell inequalities for graph states. *Phys. Rev. Lett.*, **95**, 120405.

Haagerup, Uffe and Musat, Magdalena (2011, Apr). Factorization and dilation problems for completely positive maps on von neumann algebras. *Communications in Mathematical Physics*, **303** (2), 555–594.

Hall, Michael J. W. (2011, Aug). Relaxed Bell inequalities and Kochen-Specker theorems. *Phys. Rev. A*, **84**, 022102.

Handsteiner, Johannes, Friedman, Andrew S., Rauch, Dominik, Gallicchio, Jason, Liu, Bo, Hosp, Hannes, Kofler, Johannes, Bricher, David, Fink, Matthias, Leung, Calvin, Mark, Anthony, Nguyen, Hien T., Sanders, Isabella, Steinlechner, Fabian, Ursin, Rupert, Wengerowsky, Sören, Guth, Alan H., Kaiser, David I., Scheidl, Thomas, and Zeilinger, Anton (2017, Feb). Cosmic Bell test: Measurement settings from milky way stars. *Phys. Rev. Lett.*, **118**, 060401.

Hänggi, Esther, Renner, Renato, and Wolf, Stefan (2013). The impossibility of non-signaling privacy amplification. *Theoretical Computer Science*, **486**, 27–42.

Hardy, Lucien (1992, May). Quantum mechanics, local realistic theories, and lorentz-invariant realistic theories. *Phys. Rev. Lett.*, **68**, 2981–2984.

Hardy, Lucien (1993, Sep). Nonlocality for two particles without inequalities for almost all entangled states. *Phys. Rev. Lett.*, **71**, 1665–1668.

Hardy, Lucien (2004). Quantum ontological excess baggage. *Studies in History and Philosophy of Science Part B: Studies in History and Philosophy of Modern Physics*, **35** (2), 267–276.

Hensen, B., Bernien, H., Dreau, A. E., Reiserer, A., Kalb, N., Blok, M. S., Ruitenberg, J., Vermeulen, R. F. L., Schouten, R. N., Abellan, C., Amaya, W., Pruneri, V., Mitchell, M. W., Markham, M., Twitchen, D. J., Elkouss, D., Wehner, S., Taminiau, T. H., and Hanson, R. (2015, Oct). Loophole-free Bell inequality violation using electron spins separated by 1.3 kilometres. *Nature*, **526** (7575), 682–686.

Hirsch, Flavien, Quintino, Marco Túlio, Bowles, Joseph, and Brunner, Nicolas (2013, Oct). Genuine hidden quantum nonlocality. *Phys. Rev. Lett.*, **111**, 160402.

Hirsch, Flavien, Quintino, Marco Túlio, and Brunner, Nicolas (2018, Jan). Quantum measurement incompatibility does not imply Bell nonlocality. *Phys. Rev. A*, **97**, 012129.

Hirsch, Flavien, Quintino, Marco Tulio, Vértesi, Tamas, Navascués, Miguel, and Brunner, Nicolas (2017). Better local hidden variable models for two-qubit Werner states and an upper bound on the Grothendieck constant kg(3). *Quantum*, **1**, 3.

Horodecki, R., Horodecki, P., and Horodecki, M. (1995). Violating Bell inequality by mixed spin-1/2 states: Necessary and sufficient condition. *Physics Letters A*, **200** (5), 340–344.

Horodecki, Ryszard, Horodecki, Paweł, Horodecki, Michał, and Horodecki, Karol (2009, Jun). Quantum entanglement. *Rev. Mod. Phys.*, **81**, 865–942.

Ji, Zhengfeng, Natarjan, Anand, Vidick, Thomas, Wright, John and Yuen, Henry (2020). MIP* = RE. Preprint arXiv:2001.04383.

Jones, Nick S., Linden, Noah, and Massar, Serge (2005, Apr). Extent of multiparticle quantum nonlocality. *Phys. Rev. A*, **71**, 042329.

Jones, Nick S. and Masanes, Lluís (2005, Nov). Interconversion of nonlocal correlations. *Phys. Rev. A*, **72**, 052312.

Kaniewski, Jedrzej (2016, Aug). Analytic and nearly optimal self-testing bounds for the Clauser-Horne-Shimony-Holt and Mermin inequalities. *Phys. Rev. Lett.*, **117**, 070402.

Kaniewski, Jedrzej (2017, Jun). Self-testing of binary observables based on commutation. *Phys. Rev. A*, **95**, 062323.

Kaszlikowski, Dagomir, Kwek, L. C., Chen, Jing-Ling, Żukowski, Marek, and Oh, C. H. (2002, Feb). Clauser-Horne inequality for three-state systems. *Phys. Rev. A*, **65**, 032118.

Kessler, Max and Arnon-Friedman, Rotem (2017). Device-independent randomness amplification and privatization. Preprint arXiv:1705.04148.

Khalfin, L. A. and Tsirelson, B. S. (1985). Quantum and quasi-classical analogs of Bell inequalities. In *Symposium on the Foundations of Modern Physics* (ed. P. Lathi and P. Mittelstaedt), pp. 441–460. World Scientific Publishing, Singapore.

Klyachko, Alexander A., Can, M. Ali, Binicioğlu, Sinem, and Shumovsky, Alexander S. (2008, Jul). Simple test for hidden variables in spin-1 systems. *Phys. Rev. Lett.*, **101**, 020403.

Kochen, S. and Specker, E. P. (1967). The problem of hidden variables in quantum mechanics. *Journal of Mathematics and Mechanics*, **17**, 59–87.

Koh, Dax Enshan, Hall, Michael J. W., Setiawan, Pope, James E., Marletto, Chiara, Kay, Alastair, Scarani, Valerio, and Ekert, Artur (2012, Oct). Effects of reduced measurement independence on Bell-based randomness expansion. *Phys. Rev. Lett.*, **109**, 160404.

Landau, Lawrence J. (1988, Apr). Empirical two-point correlation functions. *Foundations of Physics*, 18 (4), 449–460.

Larsson, Jan-Åke (2014). Loopholes in Bell inequality tests of local realism. *Journal of Physics A: Mathematical and Theoretical*, 47 (42), 424003.

Law, Yun Zhi, Thinh, Le Phuc, Bancal, Jean-Daniel, and Scarani, Valerio (2014). Quantum randomness extraction for various levels of characterization of the devices. *Journal of Physics A: Mathematical and Theoretical*, 47 (42), 424028.

Leggett, A. J. (2003, Oct). Nonlocal hidden-variable theories and quantum mechanics: An incompatibility theorem. *Foundations of Physics*, 33 (10), 1469–1493.

Leifer, Matthew (2014). Is the quantum state real? An extended review of ψ-ontology theorems. *Quanta*, 3 (1), 67–155.

Lin, Pei-Sheng, Rosset, Denis, Zhang, Yanbao, Bancal, Jean-Daniel, and Liang, Yeong-Cherng (2018, Mar). Device-independent point estimation from finite data and its application to device-independent property estimation. *Phys. Rev. A*, 97, 032309.

Liu, Yang, Yuan, Xiao, Li, Ming-Han, Zhang, Weijun, Zhao, Qi, Zhong, Jiaqiang, Cao, Yuan, Li, Yu-Huai, Chen, Luo-Kan, Li, Hao, Peng, Tianyi, Chen, Yu-Ao, Peng, Cheng-Zhi, Shi, Sheng-Cai, Wang, Zhen, You, Lixing, Ma, Xiongfeng, Fan, Jingyun, Zhang, Qiang, and Pan, Jian-Wei (2018, Jan). High-speed device-independent quantum random number generation without a detection loophole. *Phys. Rev. Lett.*, 120, 010503.

López-Rosa, Sheila, Xu, Zhen-Peng, and Cabello, Adán (2016, Dec). Maximum nonlocality in the (3,2,2) scenario. *Phys. Rev. A*, 94, 062121.

Maassen, Hans and Uffink, J. B. M. (1988, Mar). Generalized entropic uncertainty relations. *Phys. Rev. Lett.*, 60, 1103–1106.

Martin, A., Kaiser, F., Vernier, A., Beveratos, A., Scarani, V., and Tanzilli, S. (2013, Feb). Cross time-bin photonic entanglement for quantum key distribution. *Phys. Rev. A*, 87, 020301.

Masanes, Lluís (2003a). Necessary and sufficient condition for quantum-generated correlations. Preprint arXiv:quant-ph/0309137.

Masanes, Lluís (2003b). Tight Bell inequality for d-outcome measurements correlations. *Quantum Inf. Comput.*, 3, 345.

Masanes, Ll., Acin, A., and Gisin, N. (2006, Jan). General properties of nonsignaling theories. *Phys. Rev. A*, 73, 012112.

Masanes, Lluís, Liang, Yeong-Cherng, and Doherty, Andrew C. (2008, Mar). All bipartite entangled states display some hidden nonlocality. *Phys. Rev. Lett.*, 100, 090403.

Máttar, Alejandro, Kołodyński, Jan, Skrzypczyk, Paul, Cavalcanti, Daniel, Banaszek, Konrad, and Acín, Antonio (2020). Device-independent quantum key distribution with single-photon sources. Quantum 4, 260.

Maudlin, Tim (2011). *Quantum Non-Locality and Relativity*. Wiley-Blackwell, Chichester.

Maudlin, Tim (2014). What Bell did. *Journal of Physics A: Mathematical and Theoretical*, 47 (42), 424010.

Mayers, Dominic and Yao, Andrew (1998). Quantum cryptography with imperfect apparatus. In *Proceedings of the 39th Annual Symposium on the Foundations of Computer Science* (ed. D. Kurlander, M. Brown, and R. Rao), pp. 503–509. IEEE, New York.

Mayers, Dominic and Yao, Andrew (2004). Self testing quantum apparatus. *Quant. Inf. Comput.*, 4, 273–286.

McKague, Matthew, Yang, Tzyh Haur, and Scarani, Valerio (2012). Robust self-testing of the singlet. *Journal of Physics A: Mathematical and Theoretical*, 45 (45), 455304.

Mermin, N. D. (1981). Bringing home the atomic world: Quantum mysteries for anybody. *American Journal of Physics*, 49 (10), 940–943.

Mermin, N. David (1990*a*, Oct). Extreme quantum entanglement in a superposition of macroscopically distinct states. *Phys. Rev. Lett.*, **65**, 1838–1840.

Mermin, N. David (1990*b*). Quantum mysteries revisited. *American Journal of Physics*, **58** (8), 731–734.

Mironowicz, Piotr and Pawłowski, Marcin (2013, Sep). Robustness of quantum-randomness expansion protocols in the presence of noise. *Phys. Rev. A*, **88**, 032319.

Montina, A. (2006, Oct). Condition for any realistic theory of quantum systems. *Phys. Rev. Lett.*, **97**, 180401.

Moroder, Tobias, Bancal, Jean-Daniel, Liang, Yeong-Cherng, Hofmann, Martin, and Gühne, Otfried (2013, Jul). Device-independent entanglement quantification and related applications. *Phys. Rev. Lett.*, **111**, 030501.

Navascués, M., Cooney, T., Pérez-García, D., and Villanueva, N. (2012, Aug). A physical approach to tsirelson's problem. *Foundations of Physics*, **42** (8), 985–995.

Navascués, Miguel, Guryanova, Yelena, Hoban, Matty J., and Acín, Antonio (2015, 02). Almost quantum correlations. *Nature Communications*, **6**, 6288.

Navascués, Miguel, Pironio, Stefano, and Acín, Antonio (2007, Jan). Bounding the set of quantum correlations. *Phys. Rev. Lett.*, **98**, 010401.

Navascués, Miguel, Pironio, Stefano, and Acín, Antonio (2008). A convergent hierarchy of semidefinite programs characterizing the set of quantum correlations. *New Journal of Physics*, **10** (7), 073013.

Navascués, Miguel and Wunderlich, Harald (2010). A glance beyond the quantum model. *Proceedings of the Royal Society of London A: Mathematical, Physical and Engineering Sciences*, **466** (2115), 881–890.

Nielsen, Michael and Chuang, Isaac (2000). *Quantum Information and Computation*. Cambridge University Press, Cambridge.

Norsen, Travis (2007, Mar). Against 'realism'. *Foundations of Physics*, **37** (3), 311–340.

Oszmaniec, Michał, Guerini, Leonardo, Wittek, Peter, and Acín, Antonio (2017, Nov). Simulating positive-operator-valued measures with projective measurements. *Phys. Rev. Lett.*, **119**, 190501.

Pál, Károly F. and Vértesi, Tamás (2010, Aug). Maximal violation of a bipartite three-setting, two-outcome Bell inequality using infinite-dimensional quantum systems. *Phys. Rev. A*, **82**, 022116.

Pál, Károly F. and Vértesi, Tamás (2015, Nov). Bell inequalities violated using detectors of low efficiency. *Phys. Rev. A*, **92**, 052104.

Palazuelos, Carlos (2012, Nov). Superactivation of quantum nonlocality. *Phys. Rev. Lett.*, **109**, 190401.

Pan, Jian-Wei, Chen, Zeng-Bing, Lu, Chao-Yang, Weinfurter, Harald, Zeilinger, Anton, and Żukowski, Marek (2012, May). Multiphoton entanglement and interferometry. *Rev. Mod. Phys.*, **84**, 777–838.

Pawłowski, Marcin, Paterek, Tomasz, Kaszlikowski, Dagomir, Scarani, Valerio, Winter, Andreas, and Żukowski, Marek (2009, Oct). Information causality as a physical principle. *Nature*, **461**, 1101.

Pawłowski, Marcin and Scarani, Valerio (2016). Information causality. In *Quantum Theory: Informational Foundations and Foils* (ed. G. Chiribella and R. W. Spekkens), pp. 423–438. Springer, Dordrecht.

Pearle, Philip M. (1970, Oct). Hidden-variable example based upon data rejection. *Phys. Rev. D*, **2**, 1418–1425.

Peres, Asher (1996, Oct). Collective tests for quantum nonlocality. *Phys. Rev. A*, **54**, 2685–2689.

Pironio, Stefano (2015). Random 'choices' and the locality loophole. Preprint arXiv:1510.00248.

Pironio, Stefano, Acín, Antonio, Brunner, Nicolas, Gisin, Nicolas, Massar, Serge, and Scarani, Valerio (2009). Device-independent quantum key distribution secure against collective attacks. *New Journal of Physics*, **11** (4), 045021.

Pironio, S., Acín, A., Massar, S., de la Giroday, A. Boyer, Matsukevich, D. N., Maunz, P., Olmschenk, S., Hayes, D., Luo, L., Manning, T. A., and Monroe, C. (2010, Apr). Random numbers certified by Bell's theorem. *Nature*, **464**, 1021.

Pironio, Stefano, Bancal, Jean-Daniel, and Scarani, Valerio (2011). Extremal correlations of the tripartite no-signaling polytope. *Journal of Physics A: Mathematical and Theoretical*, **44** (6), 065303.

Pironio, Stefano and Massar, Serge (2013, Jan). Security of practical private randomness generation. *Phys. Rev. A*, **87**, 012336.

Pitowsky, Itamar (1989). From George Boole to john Bell—the Origins of Bell's inequalities. In *Bell's Theorem, Quantum Theory, and Conceptions of the Universe* (ed. M. Kafatos), pp. 37–49. Kluwer Academic, Dordrecht.

Pitowsky, Itamar and Svozil, Karl (2001, Jun). Optimal tests of quantum nonlocality. *Phys. Rev. A*, **64**, 014102.

Pivoluska, Matej and Plesch, Martin (2014). Device independent random number generation. *Acta Physica Slovaca*, **64** (6), 600–663.

Popescu, Sandu (1994, Feb). Bell's inequalities versus teleportation: What is nonlocality? *Phys. Rev. Lett.*, **72**, 797–799.

Popescu, Sandu (1995, Apr). Bell's inequalities and density matrices: Revealing "hidden" nonlocality. *Phys. Rev. Lett.*, **74**, 2619–2622.

Popescu, Sandu and Rohrlich, Daniel (1992*a*). Generic quantum nonlocality. *Physics Letters A*, **166** (5-6), 293–297.

Popescu, Sandu and Rohrlich, Daniel (1992*b*). Which states violate Bell's inequality maximally? *Physics Letters A*, **169** (6), 411–414.

Popescu, Sandu and Rohrlich, Daniel (1994). Quantum nonlocality as an axiom. *Foundations of Physics*, **24**, 379.

Portmann, Samuel, Branciard, Cyril, and Gisin, Nicolas (2012, Jul). Local content of all pure two-qubit states. *Phys. Rev. A*, **86**, 012104.

Pusey, Matthew F., Barrett, Jonathan, and Rudolph, Terry (2012, May). On the reality of the quantum state. *Nature Physics*, **8**, 475.

Pütz, Gilles and Gisin, Nicolas (2016). Measurement dependent locality. *New Journal of Physics*, **18** (5), 055006.

Pütz, Gilles, Martin, Anthony, Gisin, Nicolas, Aktas, Djeylan, Fedrici, Bruno, and Tanzilli, Sébastien (2016, Jan). Quantum nonlocality with arbitrary limited detection efficiency. *Phys. Rev. Lett.*, **116**, 010401.

Pütz, Gilles, Rosset, Denis, Barnea, Tomer Jack, Liang, Yeong-Cherng, and Gisin, Nicolas (2014, Nov). Arbitrarily small amount of measurement independence is sufficient to manifest quantum nonlocality. *Phys. Rev. Lett.*, **113**, 190402.

Rabelo, Rafael, Ho, Melvyn, Cavalcanti, Daniel, Brunner, Nicolas, and Scarani, Valerio (2011, Jul). Device-independent certification of entangled measurements. *Phys. Rev. Lett.*, **107**, 050502.

Rabelo, Rafael, Law, Yun Zhi, and Scarani, Valerio (2012, Oct). Device-independent bounds for hardy's experiment. *Phys. Rev. Lett.*, **109**, 180401.

Rai, Ashutosh, Duarte, Cristiano, Brito, Samurai, and Chaves, Rafael (2019). Into the quantum void: Geometry of the quantum set on no-signaling faces. Phys. Rev. A **99**, 032106.

Rastall, Peter (1985, Sep). Locality, Bell's theorem, and quantum mechanics. *Foundations of Physics*, **15** (9), 963–972.

Reichardt, Ben W., Unger, Falk, and Vazirani, Umesh (2013, Apr). Classical command of quantum systems. *Nature*, **496**, 456.

Renner, Renato (2007, Jul). Symmetry of large physical systems implies independence of subsystems. *Nature Physics*, **3**, 645.

Renou, Marc-Olivier, Wang, Yuyi, Boreiri, Sadra, Beigi, Salman, Gisin, Nicolas, and Brunner, Nicolas (2019). Limits on correlations in networks for quantum and no-signaling resources. Phys. Rev. Lett. 123, 070403 (2019).

Rosenfeld, Wenjamin, Burchardt, Daniel, Garthoff, Robert, Redeker, Kai, Ortegel, Norbert, Rau, Markus, and Weinfurter, Harald (2017, Jul). Event-ready Bell test using entangled atoms simultaneously closing detection and locality loopholes. *Phys. Rev. Lett.*, **119**, 010402.

Rosset, Denis, Bancal, Jean-Daniel, and Gisin, Nicolas (2014). Classifying 50 years of Bell inequalities. *Journal of Physics A: Mathematical and Theoretical*, **47** (42), 424022.

Rowe, M. A., Kielpinski, D., Meyer, V., Sackett, C. A., Itano, W. M., Monroe, C., and Wineland, D. J. (2001, Feb). Experimental violation of a Bell's inequality with efficient detection. *Nature*, **409** (6822), 791–794.

Roy, S. M. and Singh, Virendra (1991, Nov). Tests of signal locality and Einstein-Bell locality for multiparticle systems. *Phys. Rev. Lett.*, **67**, 2761–2764.

Salart, Daniel, Baas, Augustin, Branciard, Cyril, Gisin, Nicolas, and Zbinden, Hugo (2008, Aug). Testing the speed of "spooky action at a distance". *Nature*, **454**, 861.

Scarani, Valerio (2008, Apr). Local and nonlocal content of bipartite qubit and qutrit correlations. *Phys. Rev. A*, **77**, 042112.

Scarani, Valerio, Acín, Antonio, Schenck, Emmanuel, and Aspelmeyer, Markus (2005, Apr). Nonlocality of cluster states of qubits. *Phys. Rev. A*, **71**, 042325.

Scarani, Valerio, Bancal, Jean-Daniel, Suarez, Antoine, and Gisin, Nicolas (2014, May). Strong constraints on models that explain the violation of Bell inequalities with hidden superluminal influences. *Foundations of Physics*, **44** (5), 523–531.

Scarani, Valerio, Bechmann-Pasquinucci, Helle, Cerf, Nicolas J., Dušek, Miloslav, Lütkenhaus, Norbert, and Peev, Momtchil (2009, Sep). The security of practical quantum key distribution. *Rev. Mod. Phys.*, **81**, 1301–1350.

Scarani, Valerio and Gisin, Nicolas (2001). Spectral decomposition of Bell's operators for qubits. *Journal of Physics A: Mathematical and General*, **34** (30), 6043.

Scarani, Valerio and Gisin, Nicolas (2002). Superluminal influences, hidden variables, and signaling. *Physics Letters A*, **295** (4), 167–174.

Scarani, Valerio and Gisin, Nicolas (2005, Jun). Superluminal hidden communication as the underlying mechanism for quantum correlations: Constraining models. *Brazilian Journal of Physics*, **35**, 328–332.

Scarani, Valerio, Gisin, Nicolas, Brunner, Nicolas, Masanes, Lluis, Pino, Sergi, and Acín, Antonio (2006, Oct). Secrecy extraction from no-signaling correlations. *Phys. Rev. A*, **74**, 042339.

Schmied, Roman, Bancal, Jean-Daniel, Allard, Baptiste, Fadel, Matteo, Scarani, Valerio, Treutlein, Philipp, and Sangouard, Nicolas (2016). Bell correlations in a Bose-Einstein condensate. *Science*, **352** (6284), 441–444.

Schumacher, Benjamin and Westmoreland, Michael (2010). *Quantum Processes, Systems and Information*. Cambridge University Press, Cambridge.

Seevinck, Michael and Svetlichny, George (2002, Jul). Bell-type inequalities for partial separability in *n*-particle systems and quantum mechanical violations. *Phys. Rev. Lett.*, **89**, 060401.

Shalm, Lynden K., Meyer-Scott, Evan, Christensen, Bradley G., Bierhorst, Peter, Wayne, Michael A., Stevens, Martin J., Gerrits, Thomas, Glancy, Scott, Hamel, Deny R., Allman, Michael S., Coakley, Kevin J., Dyer, Shellee D., Hodge, Carson, Lita, Adriana E., Verma, Varun B., Lambrocco, Camilla, Tortorici, Edward, Migdall, Alan L., Zhang, Yanbao, Kumor, Daniel R., Farr, William H., Marsili, Francesco, Shaw, Matthew D., Stern, Jeffrey A., Abellán, Carlos, Amaya, Waldimar, Pruneri, Valerio, Jennewein, Thomas, Mitchell, Morgan W., Kwiat, Paul G., Bienfang, Joshua C., Mirin, Richard P., Knill, Emanuel, and Nam, Sae Woo (2015, Dec). Strong loophole-free test of local realism. *Phys. Rev. Lett.*, **115**, 250402.

Shen, Lijiong, Lee, Jianwei, Thinh, Le Phuc, Bancal, Jean-Daniel, Cerè, Alessandro, Lamas-Linares, Antia, Lita, Adriana, Gerrits, Thomas, Nam, Sae Woo, Scarani, Valerio, and Kurtsiefer, Christian (2018, Oct). Randomness extraction from bell violation with continuous parametric down-conversion. *Phys. Rev. Lett.*, **121**, 150402.

Shor, Peter W. (1994, Nov). Algorithms for quantum computation: Discrete logarithms and factoring. In *Proceedings 35th Annual Symposium on Foundations of Computer Science*, pp. 124–134.

Silman, J., Pironio, S., and Massar, S. (2013, Mar). Device-independent randomness generation in the presence of weak cross-talk. *Phys. Rev. Lett.*, **110**, 100504.

Silva, Ralph, Gisin, Nicolas, Guryanova, Yelena, and Popescu, Sandu (2015, Jun). Multiple observers can share the nonlocality of half of an entangled pair by using optimal weak measurements. *Phys. Rev. Lett.*, **114**, 250401.

Skrzypczyk, Paul, Brunner, Nicolas, and Popescu, Sandu (2009, Mar). Emergence of quantum correlations from nonlocality swapping. *Phys. Rev. Lett.*, **102**, 110402.

Śliwa, Cezary (2003). Symmetries of the Bell correlation inequalities. *Physics Letters A*, **317** (3), 165–168.

Slofstra, William (2016). Tsirelson's problem and an embedding theorem for groups arising from non-local games. Preprint arXiv:1606.03140.

Spekkens, R. W. (2005, May). Contextuality for preparations, transformations, and unsharp measurements. *Phys. Rev. A*, **71**, 052108.

Stefanov, André, Zbinden, Hugo, Gisin, Nicolas, and Suarez, Antoine (2002, Mar). Quantum correlations with spacelike separated beam splitters in motion: Experimental test of multisi-multaneity. *Phys. Rev. Lett.*, **88**, 120404.

Suarez, Antoine and Scarani, Valerio (1997). Does entanglement depend on the timing of the impacts at the beam-splitters? *Physics Letters A*, **232** (1), 9–14.

Summers, Stephen J. and Werner, Reinhard (1987). Bell's inequalities and quantum field theory. *Journal of Mathematical Physics*, **28** (10), 2440–2447.

Šupić, I, Coladangelo, A, Augusiak, R, and Acín, A (2018, Aug). Self-testing multipartite entangled states through projections onto two systems. *New Journal of Physics*, **20** (8), 083041.

Šupić, Ivan and Bowles, Joseph (2020). Self-testing of quantum systems: a review. Quantum **4**, 337.

Šupić, Ivan and Bowles, Joseph and Renou, Marc-Olivier and Acín, Antonio and Hoban, Matty J., Quantum networks self-test all entangled states, (2023) Nat. Phys. **19**, 670.

Svetlichny, George (1987, May). Distinguishing three-body from two-body nonseparability by a Bell-type inequality. *Phys. Rev. D*, **35**, 3066–3069.

Tan, Ernest Y.-Z., Cai, Yu, and Scarani, Valerio (2016, Sep). Measurement-dependent locality beyond independent and identically distributed runs. *Phys. Rev. A*, **94**, 032117.

Tasca, D. S., Walborn, S. P., Toscano, F., and Souto Ribeiro, P. H. (2009, Sep). Observation of tunable popescu-rohrlich correlations through postselection of a gaussian state. *Phys. Rev. A*, **80**, 030101.

Tavakoli, Armin and Cabello, Adán (2018, Mar). Quantum predictions for an unmeasured system cannot be simulated with a finite-memory classical system. *Phys. Rev. A*, **97**, 032131.

Thinh, Le Phuc, Sheridan, Lana, and Scarani, Valerio (2013, Jun). Bell tests with min-entropy sources. *Phys. Rev. A*, 87, 062121.

Tittel, W., Brendel, J., Zbinden, H., and Gisin, N. (1998, Oct). Violation of Bell inequalities by photons more than 10 km apart. *Phys. Rev. Lett.*, 81, 3563–3566.

Toner, B. F. and Bacon, D. (2003, Oct). Communication cost of simulating bell correlations. *Phys. Rev. Lett.*, 91, 187904.

Tsirel'son, B. S. (1987, Feb). Quantum analogues of the Bell inequalities. the case of two spatially separated domains. *Journal of Soviet Mathematics*, 36 (4), 557–570.

Tura, J., Augusiak, R., Sainz, A. B., Vértesi, T., Lewenstein, M., and Acín, A. (2014). Detecting nonlocality in many-body quantum states. *Science*, 344 (6189), 1256–1258.

van Dam, Wim (2013, Mar). Implausible consequences of superstrong nonlocality. *Natural Computing*, 12 (1), 9–12.

Van Himbeeck, Thomas, Woodhead, Erik, Cerf, Nicolas J., García-Patrón, Raúl, and Pironio, Stefano (2017). Semi-device-independent framework based on natural physical assumptions. *Quantum*, 1, 33.

Vértesi, T. (2008, Sep). More efficient Bell inequalities for werner states. *Phys. Rev. A*, 78, 032112.

Vértesi, Tamás and Brunner, Nicolas (2014, Nov). Disproving the peres conjecture by showing Bell nonlocality from bound entanglement. *Nat. Comm.*, 5, 5297.

Vértesi, Tamás and Pál, Károly F. (2009, Apr). Bounding the dimension of bipartite quantum systems. *Phys. Rev. A*, 79, 042106.

Wang, Yukun, Wu, Xingyao, and Scarani, Valerio (2016). All the self-testings of the singlet for two binary measurements. *New Journal of Physics*, 18 (2), 025021.

Watrous, John (2018). *The Theory of Quantum Information*. Cambridge University Press, Cambridge.

Wehner, Stephanie (2006, Feb). Tsirelson bounds for generalized clauser-horne-shimony-holt inequalities. *Phys. Rev. A*, 73, 022110.

Wehner, Stephanie, Christandl, Matthias, and Doherty, Andrew C. (2008, Dec). Lower bound on the dimension of a quantum system given measured data. *Phys. Rev. A*, 78, 062112.

Weihs, Gregor, Jennewein, Thomas, Simon, Christoph, Weinfurter, Harald, and Zeilinger, Anton (1998, Dec). Violation of Bell's inequality under strict Einstein locality conditions. *Phys. Rev. Lett.*, 81, 5039–5043.

Werner, Reinhard F. (1989, Oct). Quantum states with Einstein-Podolsky-Rosen correlations admitting a hidden-variable model. *Phys. Rev. A*, 40, 4277–4281.

Werner, Reinhard F (2014). Comment on "what Bell did". *Journal of Physics A: Mathematical and Theoretical*, 47 (42), 424011.

Werner, R. F. and Wolf, M. M. (2001a, Aug). All-multipartite Bell-correlation inequalities for two dichotomic observables per site. *Phys. Rev. A*, 64, 032112.

Werner, R. F. and Wolf, M. M. (2001b, Nov). Bell inequalities and entanglement. *Quantum Inf. Comput.*, 1, 1.

Wiseman, Howard M (2014). The two Bell's theorems of John Bell. *Journal of Physics A: Mathematical and Theoretical*, 47 (42), 424001.

Wiseman, H. M., Jones, S. J., and Doherty, A. C. (2007, Apr). Steering, entanglement, nonlocality, and the Einstein-Podolsky-Rosen paradox. *Phys. Rev. Lett.*, 98, 140402.

Wood, Christopher J., and Spekkens, Robert W. (2015). The lesson of causal discovery algorithms for quantum correlations: Causal explanations of Bell-inequality violations require fine-tuning. *New Journal of Physics*, 17 (3), 033002.

Wu, Xingyao, Bancal, Jean-Daniel, McKague, Matthew, and Scarani, Valerio (2016, Jun). Device-independent parallel self-testing of two singlets. *Phys. Rev. A*, **93**, 062121.

Yu, Sixia and Oh, C. H. (2012, Jan). State-independent proof of Kochen-Specker theorem with 13 rays. *Phys. Rev. Lett.*, **108**, 030402.

Zambrini Cruzeiro, Emanuel and Gisin, Nicolas (2019). Complete list of Bell inequalities with four binary settings. Phys. Rev. A **99**, 022104.

Zeilinger, Anton (2010). *Dance of the photons*. Farrar, Straus and Giroux, New York.

Zhang, Yanbao, Glancy, Scott, and Knill, Emanuel (2013, Nov). Efficient quantification of experimental evidence against local realism. *Phys. Rev. A*, **88**, 052119.

Zhou, Yuqian, Cai, Yu, Bancal, Jean-Daniel, Gao, Fei, and Scarani, Valerio (2017, Nov). Many-box locality. *Phys. Rev. A*, **96**, 052108.

Zohren, Stefan and Gill, Richard D. (2008, Mar). Maximal violation of the Collins-Gisin-Linden-Massar-Popescu inequality for infinite dimensional states. *Phys. Rev. Lett.*, **100**, 120406.

Żukowski, Marek and Brukner, Časlav (2002, May). Bell's theorem for general n-qubit states. *Phys. Rev. Lett.*, **88**, 210401.

Żukowski, Marek and Brukner, Časlav (2014). Quantum non-locality: It ain't necessarily so... *Journal of Physics A: Mathematical and Theoretical*, **47** (42), 424009.

Żukowski, Marek, Wieśniak, Marcin, and Laskowski, Wiesław (2016, Aug). Bell inequalities for quantum optical fields. *Phys. Rev. A*, **94**, 020102.

Index